the

First published in Great Britain in 2006
A & C Black Publishers Limited
38 Soho Square
London W1D 3HB
www.acblack.com

ISBN-10: 0-7136-6955-1
ISBN-13: 978-07136-6955-8

Published simultaneously in the USA by
University of Pennsylvania Press
3905 Spruce Street
Philadelphia, Pennsylvania 19104-4112

ISBN-13: 978-08122-1960-9
ISBN-10: 0-8122-1960-0

Illustrator: Viena Shilling
Book design by Paula McCann
Cover design by Sutchinda Rangsi Thompson
Copyedited by Rebecca Harman
Proofread by Carol Waters
Project Manager: Susan Kelly
Editorial Assistant: Sophie Page

Printed and bound in China by C&C Offset Printing Co., Ltd

the yarn book

How to understand, design and use yarn

Penny Walsh

A & C Black Publishers • London

University of Pennsylvania Press • Philadelphia

Contents

Acknowledgements

This book is the result of a series of workshops at Vauxhall City Farm, London, where the spinning group has for many years taken the art and craft of spinning to the highest levels of professionalism and creativity and I am grateful to all for their help and advice.

Grateful thanks are also to Graham Murrell and Nigel Swift, the photographers of these tricky examples of textiles, and to Vienna Shilling, the illustrator for re-imagining my small scribbles.

Thanks to Margaret Hall-Townley and Julie Graves of the Constance Howard Resource and Research Centre at Goldsmith's College for their co-operation and help in allowing me to research and photograph items of the collection, and to John McGlochlan and Peter Needham of UMIST and Eugene Nicholson of the Bradford Industrial Museum for answering my many technical questions. My thanks also to Mary Hulton who supplied vital information that filled a large gap, and to Professor Jerzy Maik of the Institute for Archaeology and Ethnology, Polish Academy of Sciences in Lodz, Poland, who allowed me to use his fascinating studies on the history of textiles.

I would like to acknowledge the generosity of various textile artists and companies in allowing me to use images of their work as illustrations: Eleanor Pritchard, Margo Selby and Yoko Hatakeyama, 'Mr M Designs', Rowan Yarns of Holmfirth Yorkshire and the Italian textile company Filpucci. I would also like to thank the legendary Nuno Corporation of Japan for images and information, and Nancy Munro of Handknitting.com, Dr Stan Swallow and Asha Thompson of 'Intelligent Textiles', and Julia Desch of the Stilereed Flock for the image of her Wensleydale Sheep.

Finally I would like to thank Roman Rappak and my husband Wojtek for their help with IT support.

Penny Walsh, 2006

Introduction

What is yarn?

Yarn is the basic unit of textile construction and all fabric except felt is made by manipulating it. By a combination of repeated twisting and stretching of a bundle of fibres into a continuous line, or by producing and twisting a continuous filament, the preliminary structures that make fabrics are created. Yarn is the elemental component from which the fabric receives its shape and character. The texture and appearance of fabric is created by the composition of the threads it is made from. The amount of twist in the fibres determines the balance and drape; the number of fibres and plies determines the thickness and weight of the fabric.

Sample of gold ribbed fabric designed and woven by Margo Selby. Photograph: Graham Murrell

Yarn has considerable continuous length and a small cross-section; in other words it is long and very thin. It is made up of fibres or filaments, usually bound together by twisting or spinning.

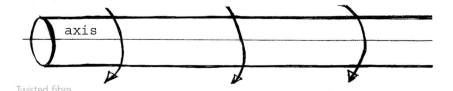

Twisted fibre

The difference between spinning and simply twisting or 'throwing' is that spinning is defined as a process that involves drawing out the threads by gripping the fibres at one end while twisting and pulling the other, whereas 'throwing' has no draw. In a spun yarn, a series of individual overlapping fibres are made into a continuous thread, bound together by pulling and rotating.

Spinning

There are two types of spinning used in yarn production:

- extrusion spinning (manufactured fibres and filament silk)
- staple spinning (natural fibres and silk waste).

In extrusion spinning, a fibre-forming substance is forced though the holes of a spinneret, beyond which it solidifies into a filament. This happens naturally in the secretion of the silk worm and has been copied by humans to create synthetically extruded yarns. Once extruded, the filaments need twisting but not stretching; they run the full length of the yarn and are smooth, strong and lustrous.

In staple spinning, short fibres are turned into yarn. A lengthening strand is produced by drawing out and combining the short lengths of the fibres, 'a material under axial strain will elongate' (the second law of thermodynamics, or put more simply 'pull it and it stretches') while twisting it under tension. The short fibres usually have to be prepared by combing or carding. This process of spinning has remained virtually unchanged since humans discovered how to twist fibres into a continuous strand,

The essential requirement of spinning is that the materials hold together, while having the ability to be taken to pieces or cut without rupturing or destroying the whole yarn. The hand or machine spinner is dealing with a material that may contain millions of individual fibres.

The identity of a yarn

The identity of an individual yarn is primarily its thickness and twist. For hand spinners and manufacturers measuring the thickness of yarn has always been a problem because any thread will compress under pressure. Yarn is mostly composed of air and the density of the fibres depends on the structure. A yarn hand spun from a woollen rolag is hollow and is 80 per cent air.

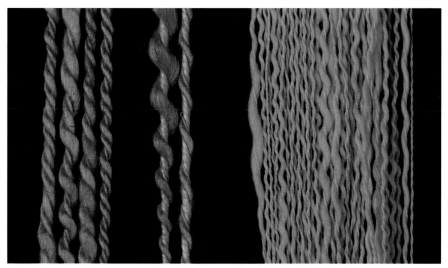

Threads formed from continuous filament silk fibres and short staple wool fibres.
Photograph: Nigel Swift

Hand spun woollen yarns

For this reason the spinning industry uses the archaic 'yarn count system' (different systems are used in different areas) a measurement based on the weight of a set length of spun thread. Filament yarn can be measured in denier traditionally used for silk, which measures the grams per 9000 metres of continuous length. These two methods have been largely replaced by the internationally agreed tex system in which the size of the yarn or filament is expressed in grams per kilometre (1000 metres), although not all yarns with the same tex have the same diameter. The newer, extra fine synthetic fibres can be measured in decitex, the weight in grammes of 10,000 metres of yarn, or in microns, one millionth of a metre. The Poisson distribution system measures the distribution of fibres within the yarn; the number of fibres in cross-section at any point along the length of a machine spun thread. Hand spinners and weavers use more traditional methods of measuring the width of a yarn; by winding it around various measures under compression.

Yarn wound around a ruler to measure width. Yarn hand spun by Pearl Allford

Twisting

The purpose of twist is to bind the fibres into a thread. Even long silk filaments will separate from each other unless they are given some twist to give axial

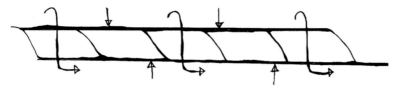

Axial (lengthways) and lateral (sideways) pressure

(lengthways) strength and lateral (sideways) pressure.

The twist level is the number of turns or twists around the axis per unit or length of thread, and all threads will have a set number of turns per centimetre or metre. As the twist level increases the yarn becomes stronger and more compact.

The twist factor defines the number of twists per centimetre in relation to the thickness of the yarn, which determines the angle the fibres lie in relation to the straight axis of the yarn.

Yarns with the same twist angle and twist factor can have different numbers of turns per centimetre depending on the thickness and strength required. Normally, finer threads have a much more acute twist angle than thick threads, but the angle can be the same, making the finer thread loose and the thicker thread tight.

Thick and thin thread with the same twist angle

The angle of twist can be measured with a protractor, and the number of twists per centimetre can be counted to get a profile of the structure of the yarn

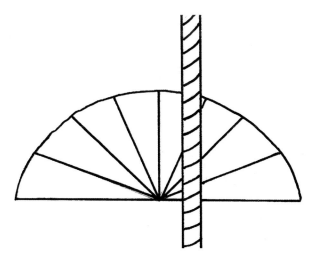

The angle of twist measured with a protractor

The strength and texture of the yarn is dependent on the twist; as the twist increases the yarn becomes stronger and more compact. For example, fine cotton warp thread has a twist factor of 40, hand-knitting wool between 20 and 30. If the twist factor is greatly increased in a fine cotton yarn with good elasticity, a crinkly fabric will result, for example, the cotton for a crepe bandage. A low twist factor and twist level used to spin fine merino will enhance the softness and lightness of the fibres.

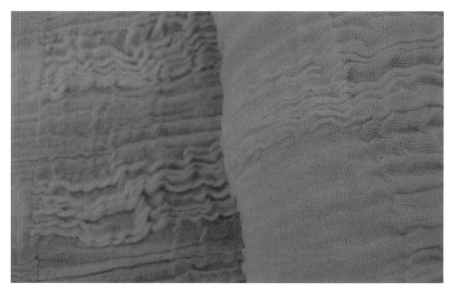

Fabric designed for the Nuno company by Reiko Sudo. A cotton fabric exploiting the the elasticity of cotton threads with differing degrees of twist

A high twist yarn will result in a firmer fabric with more crease resistance and a clearer woven structure. A fabric made with low twist yarn will be softer and more easily creased and felted.

Plying

Single threads are almost always twisted in one direction and the convention is that the fibres are rotated clockwise from the point of tension, known as 'Z'-twist

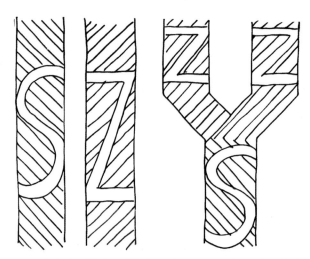

The fibres are rotated clockwise (Z), then the single threads are plied anticlockwise (S)

(following the direction of the diagonal of the letter). The single threads are then plied in the other direction, known as 'S'. This opposite twisting is known as 'twist on twist'.

The directions of spin and ply can sometimes vary with different fibres, and can be exploited for design and function.

Plying (from French plier to double over) is called 'doubling' or 'folding' in the textile industry and it adds time and expense to yarn production, so why make and ply two threads instead of one single?

The energy or torque in a twisted yarn will cause it to untwist or double back on itself, unless it is balanced or set in some way. When two single yarns are twisted together in the opposite direction to the one in which they were spun, there comes an optimum point where the two singles discharge enough energy by untwisting to grip the other thread and remain balanced, so that half the original spinning twist will remain in each single.

Balancing the torque of the thread prevents fabric from skewing out of shape.

Colour effects of plying. Photograph: Nigel Swift

Apart from balancing the twist, plying adds strength by binding the fibres together round a new axis; in fact unplied singles are generally too weak to be used for warp thread. A two ply yarn is four times stronger than a single because weak places in individual fibres are supported by fibres in the other single, and variation in strength is reduced.

The two singles lose their colour identity to make a blend. Similarly, irregularities of texture in the singles are reduced to make a uniform thread.

Plying to make yarn more bulky

Plying increases the elastic recovery of yarn, and yarn can be made more compact or bulky by plying. Lustre can also be increased and decorative constructions added.

Any number of singles can be plied together at once for added strength or fancy constructions; 2 ply, 3 ply, and so on. Sets of two or more plied yarns can be re-plied or 'cabled' together in different configurations. Such 'multi-plying' is a construction often used to increase warmth.

ONE

A History of Spinning

The history of spinning yarn stretches back to Neolithic times when humans discovered that twisting fibres into strong supple threads made better clothing than skin or hide. The construction of the yarn that is spun to make into textiles has remained broadly the same ever since. The evolution of the spinning tools found by archaeologists has provided them with a major source of cultural evidence about prehistoric societies and ancient trade routes. The history of spinning gives an insight into the social and economic history, art and costume of all the cultures of the past.

Hittite lady with drop spindle. From relief in eastern Anatolia c.1500 BC

The production of spun yarn for textiles was a vast economy involving thousands of people throughout the world up to the Industrial Revolution. Until the mid 18th century, all yarn, from lace to sack cloth, was spun by the hands of an individual spinner, and although replaced by machine spinning in western Europe in the 19th century, hand spinning is still extensively practised in many societies today. The history of spinning is obscure and mostly unrecorded. Historians have suggested that it was devalued because often (although not always) it was a woman's occupation, alternatively it may be due to the fierce secrecy surrounding the divulging of trade secrets of any part of the textile industry.

Yarn in ancient textiles

The earliest evidence of yarn comes from a fragment unearthed at Catal Huyuk in Turkey. It is a smooth bast fibre thread, as fine as a sewing thread and carbon dated as 8000 years old. In 1953 a perfect cast of spun threads, found in the Lasceaux caves in France provided evidence of the production of textiles by the Stone Age cultures of the Mediterranean area. Weights and spindles from 4000 BC have been unearthed near Damascus, and copper spindles from 3000 BC found in other parts of modern Iraq may be the high whorl spindles shown in some of the Hittite reliefs. The first samples of twisted linen thread have been discovered in Egyptian burial chambers from about the same time.

Egypt

Textile remains from ancient Egypt have been preserved in the dry climate of the desert and they show an extraordinary skill in spinning. Linen was spun as fine as gossamer and was marvelled at by other peoples. The 'fine linen of Egypt' is mentioned in the Old Testament book of Proverbs, and Greek writers recorded their envy of the sheer fabrics woven so finely that they were set at over 100 threads per centimetre. Flax depletes the soil of nutrients, and for this reason the Nile Valley, which floods annually, is the ideal place to grow it. Linen was considered a clean fibre (it absorbs moisture) and was used for mummy wraps because it resists insects. The traditional wrapped waist cloth or pleated linen kilt was traditional temple clothing in Pharaonic Egypt; wool was considered unclean. Tomb paintings show all the processes in the mass production of linen thread. Paintings in the tomb of Khnumhotep near Memphis, dated around 2430 BC, show long lines of slaves retting the flax in a river, together with male and female spinners making a roving of the longest linen fibres, rolling it into a ball, putting it into a large-mouthed jar, and spinning with hooked spindles.

The tomb paintings show mysterious techniques of spinning with two spindles at once which are now unknown. The spinners appear to twist the threads along the thigh and over the knee, and wind them on to the spindle by twisting the spindle upwards. The spun threads have been analysed as three super-fine strands of overlapping fibres, twisted together so that the overlaps of each come at different intervals to maintain strength. Egyptian linen has a high twist angle and consistent anticlockwise spin direction and it

Painting in the tomb of Knumhotep, 2430 BC. Spindle spinning linen. Left to right: supported spindles rolled on thigh; twisting spindle using long stick; 2 ply from jars behind the back; 4 ply using two spindles (technique now unknown)

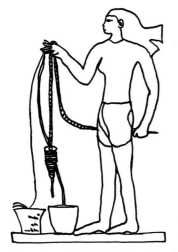

Egyptian spinners twisted the linen thread from prepared overlapping rovings

was bleached to dazzling whiteness by exposure to the sun for 8 weeks.

The expertise of the Egyptian spinners was passed on to the ancient Hebrews who used linen thread to make the shawls worn by the Hebrew priests, and to the Assyrians and Phoenicians who used spun linen for the huge sails of their ships from 1200 BC.

China

The silk industry originated in China before 2600 BC. According to legend, the wife of the Yellow Emperor, Lady His-Ling, was sitting under a mulberry tree when a silk worm dropped into her cup of hot tea. She was intrigued by the fine thread that floated away from the cocoon, and had it woven into ribbons. Silk throwing and twisting became a great industry in China from 2500 BC. During the Han dynasty, from 200 BC to AD 200, the Silk Road developed, taking the precious silk thread westward in caravans crossing the vast desolate plains of central Asia. From Peking the routes crossed the Taklamakan desert, then went either north to Kashgar and Samarkand, or south through Afghanistan and Persia, joining in Merv in Persia and on to Baghdad, Palmyra and Constantinople. From here the distribution of silk thread around the

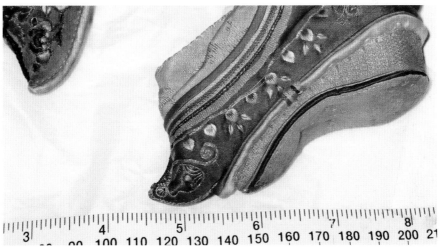

Chinese silk shoe for bound foot, embroidered with coloured silk and gold thread. From the Constance Howard collection at Goldsmith's College London

Mediterranean and to western Europe was controlled for over 1000 years.
The Chinese exported the silk, but not the knowledge, and managed to keep
the secret for a long time. However, in 140 BC a Chinese princess marrying
into the Korean royal family took silk worms and the secret of silk with her.
By AD 195 the secret had reached Japan, and in AD 550 two Christian
monks returned from China to the Emperor Justinian in Byzantium, smuggling
silk worms concealed in their bamboo walking sticks. It still took centuries
before other nations acquired the expertise of the Chinese in silk throwing
and spinning a silk warp thread strong enough to weave complex silk fabrics
on a draw loom.

Embroidery on Chinese robe using silk floss thread on silk damask background. Object no
2004.4 from The Horniman Museum and Gardens

Greece and Rome

In ancient Greece, spinning was the subject of myths and stories of the gods.
Expertise in spinning was highly admired and it is depicted on many Greek
vases. The Greeks probably acquired their expertise from the Egyptians, and
when Egypt became a province of the Roman empire in 30 BC, the Romans
took over the linen industry and set up factories round the Mediterranean.

Roman textiles were made of linen, silk and other fibres, but as spinners of
wool and breeders of specialised wool sheep, the Romans made advances in
textile manufacture that still affects our woollen clothing today. Sheep's wool

was an important fibre for the Romans; temple friezes depict small neat sheep, and the formal garments of a Roman citizen were made of wool. The toga could use 8 metres of fine wool and the short tunica was made from either white or coloured woven woollen yarn. Pliny tells us that the earlier sheep were brown or grey, and white was successfully bred to be dominant over time. Around AD 79 the inhabitants of Pompeii achieved particular skill with wool. The textile industry was one of the leading activities of the city portrayed in murals and evidenced in the spindles found in ash following the eruption of Vesuvius.

We know from Pliny that the Romans sheared and scoured fleeces and prepared them by combing, processes that changed little until the Industrial Revolution. The Roman spinners, like the Egyptians, used a distaff and a drop spindle without a whorl.

The Roman period in Britain

A writing tablet from the upper Thames Valley from AD 200 shows a Roman soldier's request for warm woollen socks. Perhaps because of the climate in Britain, the Romans organised and improved the British sheep. They thought the British sheep small and scruffy and introduced their famous white Merinos. They bred them with the native descendants of the Iron Age 'Soay', an indigenous small sheep, the descendants of which still live in the Scottish Islands. The resulting fine soft wool was used to make complex yarns and fabrics, shown by the multiple spindles of bone, wood and even jet found in Roman/British sites. Efforts by the Romans to breed the sort of soft lofty wool they liked gave Britain a variety of fine fleece-producing sheep breeds, which gave wool a uniquely important place in British history.

12 sets of 4 singles, alternating S and Z twist, Z plied into thick corded thread. Reproduced by permission of Prof. J. Maik.

The Roman period in northern Europe

Just as the dry conditions of Egypt preserved flax and linen, the wet bogs of northern Europe have preserved some of the most intricate and fascinating examples of ancient yarn spinning. Hair and wool survive well in the anaerobic bacteria-free conditions of certain peat bogs, and research by a Polish archaeologist into the structure of textiles from the Roman tombs of 1st–3rd-century Pomerania are of extreme importance to anyone with an interest in spinning and yarn. The wool fibres in the remains of textiles

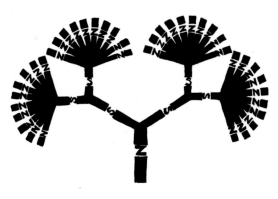

Complex cable made up of 40 singles.
Reproduced by permission of Prof. J. Maik.

in these graves were found to be of a much higher quality than later European wools of the early Middle Ages and finer than the best New Zealand Merino of the present day. The threads have an even twist and the spinning is balanced, the quality of the yarn classified as 'remarkable'. Multiple 'S'-spin singles are 'Z'-plied and 'S'-cabled into yarn constructions of great complexity. The cloaks and tunics that were woven from these yarns were intended to keep out the cold and damp of coastal northern Europe; the multiple wool plies would trap air and insulation in the threads.

A remaining mystery for spinners is that although many spinning spindles have been found from this period, no early thread container needed to manage the increasingly complex threads while plying has ever been found.

The early Middle Ages

Byzantium

From the 5th to the 14th century, the Byzantine Empire was the centre of silk yarn production. Having discovered how to grow mulberry trees and raise silk worms, although still importing great quantities from China along the Silk Road, the Byzantine silk spinners and weavers created brocades and damasks in double and compound silk weaves using techniques learned from Persian and Syrian craftsmen.

Controlling the Silk Road and distribution of silk yarn to the West, the royal workshops became rich and powerful, and five silk guilds carefully regulated every aspect of production. Byzantine spun silk yarn and twisted filament silk were woven into dalmatics and chasubles, robes for priests to wear at mass, which were exported all over the Christian world. Figured damasks were woven in complementary sets of silk threads, tightly twisted for the warp which was lifted by a complicated system of harnesses. The spinning of silk organzine (tightly twisted warp thread) was a secret learned by the Byzantine craftsmen from Sogdian and Sassanian weavers, and so closely guarded that it only reached Britain in 1732.

Example of figured weave in sets of silk filament, with tightly twisted organzine silk warp

Motifs such as griffins, lions, eagles and mythical 'senmurvs' with no overt religious imagery, were set against a patterned background. By the golden age of silk weaving in the 10th century, visitors to Byzantium commented that the ordinary people of the city wore silk robes, just to go about their daily business, and even the horses were covered with silken cloths. When Constantinople was sacked in 1420, the Byzantine silk craftsmen scattered, to Italy, Sicily and Moorish Spain.

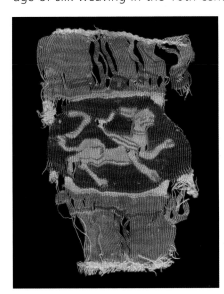

Fragment of tapestry from Fatimid period. Tiraz c.1050. Linen warp silk single twisted filament weft

Viking wool

9th to 11th century textiles have survived in York because of the moisture and low levels of oxygen in the soil there. York was an important Anglo-Scandinavian town at this time and the archaeologists who excavated the sites made extensive studies of the textile fragments. The woollen yarns appear to be spun from heavy black and white wool from long-haired north European sheep.

Much coarser than the earlier Roman wool, there is evidence of mixing the wools or breeding the sheep as time goes by. Detailed studies have shown that the wool was teased out by hand and combed, using iron-toothed Viking combs known from Norway; the softer wool was brushed with teasel thistles.

The yarns are used singly, 'Z'-spun, in weaving with a low twist angle. Wool or linen sewing thread is 2-ply 'Z', 'S', with a 20/50 degree angle. A great number of spindles and whorls have been found in the York excavations and they are the north European type, the whorls flat on one side and hemispheric in the other (bun shaped). Weighing 50-55 g, they would have been suitable for spinning heavy wool.

Contemporary spindles showing bun-shaped whorls. Photograph: Nigel Swift

The Medieval period

English early Medieval embroidery

Anglo-Saxon embroiderers used worsted thread for embroidery; the local fleeces dyed well and had a long lustrous staple. The dispute over the location of the embroidery of the Bayeux tapestry could possibly be resolved as England rather than France, due to the long wool worsted of the embroidery threads. By the 10th century, silk thread was being imported from Byzantium, some of it coming from China via the Silk Road.

Using the varied coloured silk threads, both high twist angle and softer thrown and stranded silk floss with the addition of gold threads, tinsel strips and some vegetal fibres, English embroidery became known and sought after by the wealthiest churches and courts in Europe. From the 12th to the 15th century English embroidery, known as Opus Anglicanum was famous all over Europe for its high workmanship and beauty.

In 1271 the altar front for Westminster Abbey took four nuns four years to embroider, and by 1295 there were 113 items of English embroidery in the Vatican inventory. Ecclesiastic and coronation mantles and robes were embroidered in great numbers, some in independent workshops staffed by male embroiderers who worked a seven year apprenticeship, and some in monasteries and convents. The spinners, unlike the embroiderers, did not have a guild (embroiderers had a guild from 1369) although there was an informal group of 'silkwomen'. Although the silk thread would have arrived in thrown

Cope in the style called Opus Anglicanum 1300-1320. Called the Syon Cope after the convent of Syon in Middlesex. V&A images

filaments, the fine workmanship visible in the variety of linen and silk threads, wrapped and twisted for specific areas of couched embroidery, means that the threads must have been custom made by skilled spinners.

Medieval wool

Throughout the 12th and 13th centuries wool from England was vital to the economy of Europe. The Middle Ages are known as the 'Age of the Golden Fleece' in England and the Wool Weavers are the oldest London Livery Company. By 1100, Cistercian monks were rearing large flocks of wool sheep, and exporting wool to Flanders and Italy where it was spun and woven. The famous Medieval fine wool had been selectively bred from Roman Merino to give strong soft fleeces with little or no kemp. The Romney, Ryland, Cotswold and Portland breeds had white fleeces with a long 10-15 cm staple length.

The export of wool to the great merchants of Ypres, Ghent and Bruges paid a huge amount of duty to the crown; in fact in 1191 Richard Ist's ransom was paid by wool revenue. By 1250 the merchants in London and Southampton were exporting hundreds of sacks of English wool every week. From the 12th to the 14th century there was an insatiable Flemish demand for English wool because the Flemish entrepreneurial class had the capital to buy raw materials, and they could fund the development of an important technological advance. The spinning wheel was first recorded in Europe at some time between 1350 and 1400, and was refined by the Flemish and Dutch spinners. Until this time all the yarn for all wool textiles had to be spun on the drop spindle.

The spindle wheel, turned by hand. Twisting and winding on are two processes

The 14th century European wheel was turned by hand and had no winding on mechanism, but was much more efficient than the spindle for producing large quantities of wool. Two other technical advances, the fulling mill and improvements to the loom, meant that the Flemish economy was dependent on English wool. With Italian and Hanseatic merchants competing for a share of the trade, England was able to use wool to exert political pressure. Although the European wool industry was organised on an international basis, so that fleece, yarn and fabric were sometimes transported to different countries for each process, each area had its own specialised spinning processes. The Flemish weavers favoured a strongly twisted wool, to give the heavy fabric shown in the sharp folds of the clothes in paintings by Robert Campin and other Flemish masters. The fine fleece of the Cotswold sheep was combed and lightly spun to make Cotswold blankets.

By the 14th century most European craftsmen were in guilds ruled by charters. Woolworkers were divided into several guilds: 'carders' or 'scribblers' (who prepared the wool for spinning on a table topped with wire teeth called a scribbling horse), weavers, fullers, finishers and pressers, but wool spinners were outside the Medieval guild system. Dependent on large quantities of spun wool to maintain the looms (it takes six spinners to supply one weaver) the merchants probably bound spinners to supply them with yarn by delivering and collecting the wool. Historian Heather Swanson has found that spinners were paid 19 d for 26 lb of spun thread in Colchester in the 14th century.

Trained in the convents, many women must have spun yarn full or part time to supply the weavers of each town. Girls were placed as spinning apprentices at 7 or 8 years old, often under the charge of the Weaver's Guild master's wife. Some guilds allowed women to join, if they were widowed. The records of the Florentine Wool Guild (*Arte della Lana*) show that in Italy from the 10th century, novices and orphans were taught to spin wool in convents from the age of six. By the 12th century a special elastic (possibly crepe) woollen thread was spun by these spinning schools for the stockings which were a speciality of the city.

Tapestry wool

The 14th and 15th centuries were the golden age of European tapestry; huge hangings were woven to cover great expanses of wall. In many cases the yarn used to make these large pieces of weaving has kept its strength and lustre for nearly 700 years. Unlike embroidery, tapestry has no support fabric but is created by the weft threads; the threads make the picture but also make the structure.

Spinning the warp and weft yarns for tapestry weaving required precision and experience; the yarn had to withstand lengthy dyeing processes, and

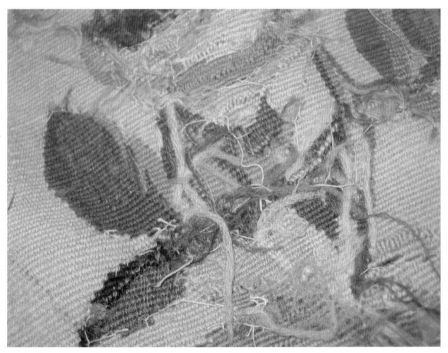

Back of a tapestry showing thick weft threads and fine linen sewing threads. Constance Howard collection, Goldsmith's College

'batting down', the method of beating the weft threads down so closely on to the warp that no warp threads are visible in the finished fabric.

Only the fleece of certain sheep is strong enough to make the finest tapestries, and larger long haired sheep were beginning to be bred in England for stronger threads. Thousands of sacks of Cotswold and north Leicester and Portland fleeces were imported by the ateliers of Paris (from 1295) and Arras from 1313. By 1420 the Duke of Burgundy was the most prominent patron of the Arras ateliers and the owner of some of the finest tapestries. The sophistication of the Mille Fleur designs he favoured, featuring an intricate floral background, required ever increasing levels of skills from the spinners and weavers. The famous Lady with Unicorn series was possibly woven for the Duke in a tapestry factory in Brussels that employed 500 people. It shows figures and exotic animals against a fantastically decorated ground in brilliant colours.

Wool, linen and silk tapestry, Story of the War of Troy, woven at Tournai in 1475. V&A images

Combed from warm fleece with heated wool combs, the wool for tapestry weft was always combed in the grease, and scoured as yarn. The combed long hair wool was probably spun from a distaff to maintain the lustre of the parallel fibres, and to ensure continuity and as few joins as possible.

Most of the threads in the tapestries of this time were a uniform worsted spun 2-ply, but it is possible to see in, for example, the Devonshire tapestries in the Victoria & Albert Museum, London, areas of texture made by using whipped singles.

Spinning long hair tapestry wool from a distaff to maintain parallel fibres

Tapestry warp had to be made of high twist non-extendable threads. Up to 1400 warps were made of heavier wool than the weft spun to a higher twist. Later tapestries use a linen warp multiplied up to 9 fold. The linen singles were 'S'-spun and 'S'-plied in multiples of odd numbers, to give a smooth cylindrical yarn. The threads were then passed through a flame to remove the anterior hairs which would eventually weaken the yarn during batting down.

By 1670 the Gobelins tapestry factory had been set up in Paris and has continued until the present day. Influenced by the workshops of Brussels, the tapestries showed an enormous increase in colour and a more painterly style, using lighter, finer wools and silk.

Multiple ply warp linen and high twist long hair tapestry weft yarn

The Renaissance

Renaissance silk

The woven silks of the Italian Renaissance are familiar from the costumes worn by the figures in the famous frescoes and paintings of the period. However, the production of the beautiful silk brocades and damask velvets of 14th- and 15th-century Italy were characterised by fanatical secrecy which protected the method of spinning fine high-twist silk-warp thread.

Techniques of silk weaving had spread along the Silk Road to Byzantium and via the Mediterranean to the city states of Italy. By the 12th century, Lucca had become a supplier of silk fabrics to the newly rich Italian bankers and financiers, and employed offshore factories in Greece and North Africa to throw and spin the vast quantities of silk thread required. Once the silk thread was reeled off the cocoons it had to be doubled and twisted by hand, in a system similar to a rope walk. At some point in the 13th century a water driven silk thrower was set up in Lucca. This circular machine contained an inner cage activated by a central vertical shaft, which caused the spindles to revolve. Effective spying meant that by the mid-14th century, Florence, Bologna and Venice had water-driven silk-throwing machines and by the 1490s Leonardo Da Vinci was experimenting in Milan with improvements to make them more efficient and to throw silk of different qualities. These improvements, along with innovations in draw-loom weaving and gold thread

Fragment of Italian 16th-century silk brocade. Constance Howard collection, Goldsmith's College

production, (all in total secrecy) meant that some of the most lavish silks and velvets in European history were produced, ensuring that Italian textiles remained paramount until the 18th century.

The Lucchese silks were woven using a heavy, loosely doubled silk weft. By the 15th century they were weaving stiffer, heavier silk damasks using the new multi-filament silk thread. The Venetians, nearest to the influence and imports from Byzantium, perfected the production of thick and thin silk thread for velvet and patterned velvet weaving, using up to five warps of different coloured silk threads. After the fall of Byzantium in 1453, skilled velvet and silk weavers fled to Venice, and by the 1460s heavy silk velvet brocades and lampas, with areas of relief picked out by differently twisted yarns and glittering gold *filé*, dyed in glowing colours were being produced in Florence, Venice, Genoa and Lucca, and were admired throughout the world.

Flax

From the 9th century, when Charlemagne encouraged flax spinning, Flanders and northern France became the centre of the Medieval linen industry. By the 13th century, the custom of spinning flax anticlockwise and plying in the same direction had been established. There were huge bleaching works at Haarlem and Bruges where linen was bleached for up to eight weeks, and the famous Toile de Rheims was an especially fine and highly esteemed linen damask used by the nobility and for liturgical vestments. Until 1780, when Arkwright constructed the cotton-spinning Jenny and linen fell out of favour, it was the

Flax preparation

fabric that everybody wore. Collars, caps, aprons, undergarments, bedding and all medical applications were made from linen, as unlike wool, it could be successfully washed and pressed. Linen thread was used for sewing all over Europe and as the core for gold thread. Russian and Scandinavian flax were imported by late Medieval times to supplement the local supply to the Flemish spinners.

Flax is a difficult fibre to spin because of its length and the complicated preparation process. A distaff would have been used to control the prepared fibres which are drawn off it in a 'line' of flax .

Spinning flax from a distaff

Until the flyer wheel was invented in the second half of the 16th century, a heavy whorl drop spindle would have been used to spin the fibres dampened with flax jelly, as finely as possible. Even with the specialised flax wheel of the 17th century, making a perfectly smooth fine linen thread was a skilled job, and in the spinning schools of Saxony and Bohemia (set up to give the poor a skill) girls were taught to spin linen from 6 years old.

Contemporary accounts describe 200 children spinning in a room supervised by a master in a pulpit in the middle with a stick to tap the lazy spinner. In another room an endless supply of distaffs were prepared and the spinning wheels kept in constant motion. In the late 16th century, flax spinners fled to England and Scotland from Holland bringing their skill and machinery with them. Flax spinning schools were set up in workhouses in Scotland and Ireland, and poor children taught to spin flax.

18th century French and English silk thread

France

Silk workers came to France from Sicily, Italy and Spain at the end of the 15th century, but the Italians kept their secret of spinning silk warp yarn until much later. They exported the strong, fine organzine thread required for the warp to weave the best quality silk cloth all over Europe at great profit.

In the 1480s Louis X1 of France had decided to encourage the French silk industry, and had brought Italian silk weavers to Lyons and Tours. By the end of the 17th century the Lyonnais could compete with the Italians. After following the Italian Renaissance designs of large classical patterns, the French began to weave smaller scale designs from finer threads, incorporating spots and stripes, constantly inventing new designs and thereby creating the concept of fashion. By 1667 manufacturers were bringing out patterns every year and Italian textiles could not keep up with the changes in taste or the amounts demanded.

The French manufacturers not only created new designs, but perfected improvements to the draw loom and new ways of spinning and preparing silk to give different textures and surface detail. Lustrings were highly shiny silk threads, stretched, heated and sometimes coated with beer; gimp was

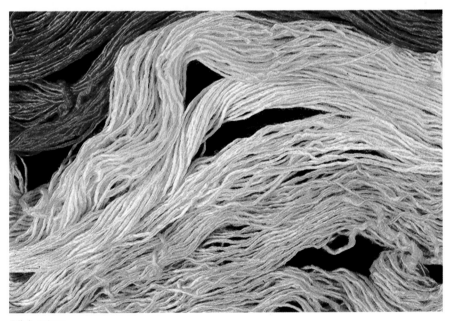

Silk lustrings

made from thicker silk coiled around a finer core thread; padusoi was tight a silk single wrapped around a linen thread; and chenille (French, caterpillar) a tufted silk four-ply thread. Multiple-ply crepe threads were used to weave several fabrics, Crepe de Chine, Crepe Marocain, Crepe Georgette (chiffon) with elastic properties and a crisp surface from the spinning technique.

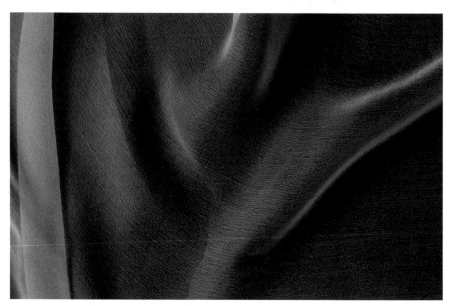

Georgette woven from crepe threads. Photograph: Nigel Swift

England

Late 16th and 17th century religious persecution meant that silk workers came to Britain in great numbers; 5000 Huguenot refugees came in 1685 following the revocation of the Edict of Nantes. They brought with them the skill to make lustrings, alamodes (thin light glossy silk) and fashionable patterned silk fabrics. By the late 17th century, in Spitalfields just outside the City of London, French, Flemish and newly trained English weavers were weaving the intricately patterned silk cloth that was in so much demand. But although the requirement for silk yarn had rocketed during the 16th century to supply the weavers, English throwing methods were unable to produce organzine for warp; they could only spin the softer 'tram' for the weft, and still relied on imports of organzine from Italy.

To produce the rope-like thread needed for organzine, single gossamer silk threads have to be twisted at 80 turns per 25 mm round a silk core, then doubled in the opposite direction in a method similar to the doubling of rope; it can then be thrown two, three or four more times. Silk could be thrown by hand, using a wheel somewhat like a ship's wheel, and a 'piecer' (often a child) who would run to the end of the silk walk and wrap the threads around

Model of Piedmontese silk throwing mill.
Silk Museum, Macclesfield, UK

a gate to double them (this procedure was still used in Leek, Staffordshire until 1945). There was also the mysterious 'Double Dutch' throwing machine, about which almost nothing is known.

In 1717 the technique of spinning silk warp thread in England was solved in a dramatic way. An English man called John Lombe from Derbyshire disguised himself as an Italian labourer and spent two years learning the secret of the Piedmontese throwing mill in northern Italy. In 1720 he returned and set up a factory in Derbyshire, which by 1722 had perfected silk organzine of the right quality.

The Piedmontese throwing mills were two stories high, and 3-4 metres in diameter, so the mills had to be constructed around them. A vertical shaft driven by water caused the inner cage to revolve, winding the filaments round the spindles.

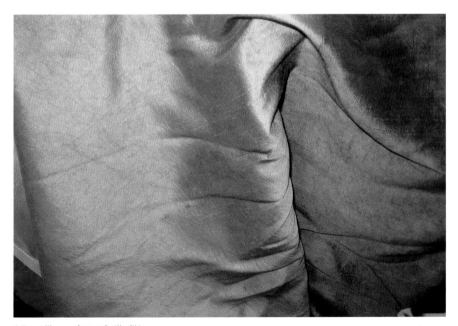

Mirror-like surface of silk *filé*

By the 1740s Warners had set up in Spitalfields and were weaving large-scale silk damasks for Syon House in Chiswick, using a new type of silk file thread. This was a *frisé* spun onto a silk core at an angle to catch the light and give a mirror-like surface. Complex patchworks of weaves in differently reflecting silk threads became popular for dress and furnishing fabrics in the second half of the 18th century.

19th century revolution in spinning

By the last quarter of the 18th century the textile industry was technologically archaic; spinning had remained unchanged for more than 300 years and fibre preparation had remained the same since antiquity. Developments in spinning technology, originally in the cotton industry, freed up the bottleneck between the fibre and the weaver, spreading to textile manufacturing and ultimately providing the impetus for the technological changes of the Industrial Revolution. In Britain, hand spinning had ceased to be part of the textile industry by the 1820s.

Cotton

In India cotton had been worn since 2000 BC. Ancient scriptures from the Indus Valley record that Indians clothed themselves exclusively in cotton, cultivating different varieties in different areas. 'Bowing' (literally tapping the seed pods with a tool like a violin bow) removed the fibres from the seeds, and spinning the fine threads by hand was a laborious and skilled process that had spread from Asia. By the 18th century, Europeans had realised the comfort of wearing cotton fabric.

References to cotton spinning in Lancashire go back to the 1600s, perhaps introduced by Flemish refugees. Cotton yarn was spun on a hand-turned wheel, a two stage operation, as winding on and twisting were not simultaneous. By the mid 18th century there was a rapidly rising demand for cotton and the weaver's output had been greatly increased by the invention of the fly-shuttle and the Dutch loom. With the invention of the carding cylinder in 1748, the pressure to spin several threads at once led to competitions between inventors and the offering of prizes by the Royal Society of Arts. At last in the 1760s two inventions came along almost at once.

In 1764 Hargreaves patented his spinning Jenny, which draws the thread out just as a hand spinner does, by the carriage pulling away from the spindles. The rotation of the spindles twists the roving just as a spinning wheel does, and on the return movement of the carriage the rotating spindles wind the yarn around themselves. In 1769 Richard Arkwright patented the water frame. This mechanism drew out the roving, not by a moving carriage, but by three sets of rollers, each going a little faster than the previous set and stretching the roving between them before flyers wound it on to bobbins.

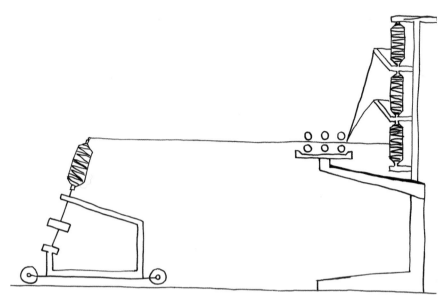

Hargreaves' spinning Jenny

Arkwright's spinning frame. Manchester Museum of Science and Industry

However, the Jenny spun soft weak thread that was not satisfactory for warp yarn, and the water frame spun thick coarse thread. The Indian dress fabrics and Palimpores admired for their softness and lightness had to be copied in Fustian which had a cotton weft and worsted warp.

In 1770 Samuel Crompton patented the mule, which spun a fine strong thread suitable even for the warp of muslins. The spindles were on a moving carriage, drawing out the thread and winding it on in one movement, and the mule became the prototype for spinning machines still in use today. It still took until the 1830s for the mule to be fully water or steam powered so that one spinner could operate a frame of 1200 spindles. The final step in the development of mechanised yarn production came from an American, John Thorpe, who in 1828 substituted a ring for the flyer, which had been modelled on a spinning wheel flyer, but which caused the spindles of a machine placed vertically instead of horizontally, to wobble.

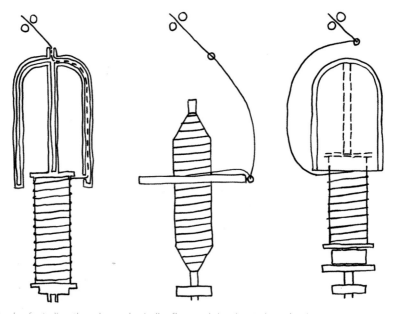

Methods of winding thread round spindle, flyer and ring (centre) mechanisms

By the 1870s the carding and drawing processes had been mechanised, and the spinning process improved so that yarn could be wound round the spindle at 6000 revolutions per minute.

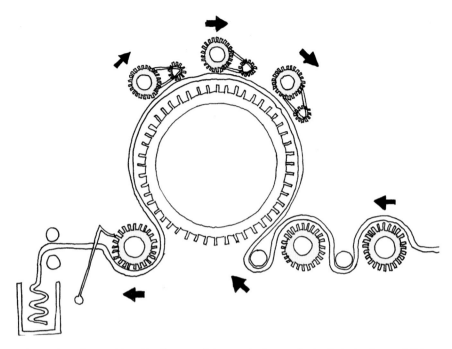

Mechanised carding machine. The fibres are fed on to the large spiked cylinder by the small 'licker in' wheel, the doffer cylinder removes the fibres and the filmy web goes into a can to be 'drawn'.

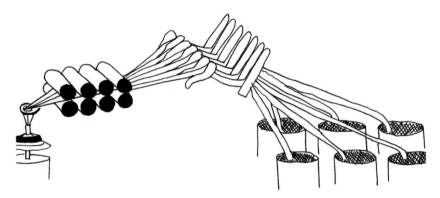

Drawing. Each of the sets of rollers runs successively faster than the previous set. The 'spoons' provide support for the slivers of fibres

Woollen and worsted yarn

Wool

From the 12th to the 19th century, wool manufacturing continued to hold a unique place in British history. Around 1600, the domestic spinning system replaced the guilds as merchants and 'broggers' began to manage the manufacture and distribution of cloth. Because of the decline of the guild system, wool spinning moved away from towns and into rural areas. After shearing, the bundles of fleece were washed in water and urine and rinsed in the running water of a stream, carded using hand carders or scribbling tables and spun by hand, before being collected by the merchant's agent. Workers owned their own spindle wheels and carding systems.

By the end of the 17th century some wool weaving and spinning was gathered into early factories in rural areas, with the spinning gallery on the top floor and the hand looms on the ground. During the 1820s spinning shops were built into the ground floor of mills, housing hand and Jenny spinning machines, together with scribbling or carding machines (a spinning shop can still be identified at Armly Mills near Leeds).

Sheep breeds such as the Shropshire, Oxford Down and Dorset Horn were developed by 18th century farmers such as Robert Bakewell. These breeds grew big fleeces of up to 7 lb on the lush pastures of the best English farmland, and supplied the growing demand for fine wool yarn for the sober suiting increasingly worn by men.

The training of spinners continued to begin at an early age, either in the family, or in institutions. In York in 1784, 30 destitute pupils are recorded as learning to spin wool from the age of 5 years. The spinning school was not closed until 1858, when its assets were donated to York Greycoats school.

During the 1770s and 1780s the wool industry began to adapt the cotton spinning machines and could machine spin a soft, strong, wool yarn using a mule. Carding processes were mechanised by the 1820s, however, it took until the early 20th century for ring spinning to be fully adapted for the rapid spinning and plying of wool. Pre-industrial textile production had involved all areas of England, but the water-powered woollen mills had gradually concentrated the spinning industry in Yorkshire, where there were several major fast flowing rivers and an abundance of fleece.

Worsted

The difference between woollen and worsted yarn had existed since the Middle Ages. In 1587 Mary Queen of Scots wore white worsted stockings called 'Guernseys' to her execution. In the 16th century Dutch spinners brought the mysterious skills of worsted combing and spinning to East Anglia, using the fine long fleece of the Norfolk Horn. The Norfolk village of Wearsted supposedly gave worsted its name. Originally the wool combing was done by women, but, in the 17th century, the combs became much heavier with long metal teeth and were heated over a stove in a huge comb pot to soften the lanolin and wax on the fleece. It then became a man's occupation, accompanied by vast quantities of beer. One comb was attached to a post and the other swung against it, drawing the fibres into a long sliver, which passed through a 'diss' or plate with a hole in, to keep it even.

Worsted carding in the 15th century. The combs are heated in a pot; one is fixed to a post to comb out the sliver

Huge metal wool combs from the 19th century. Bradford Industrial Museum

Worsted combing was the last hand process to be mechanised. The Reverend Edmund Cartwright patented the first mechanical wool comb in 1794, called, because of its circular design 'Big Ben'. The attempt to copy the actions of a hand comber had limited success, and not until the 'Noble' machine of 1856 did hand wool combing become extinct in the textile industry.

Mechanised wool combing. Bradford Industrial Museum

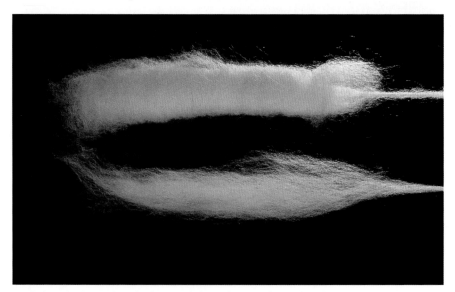

Carded wool rolag and combed worsted sliver. Photograph: Nigel Swift

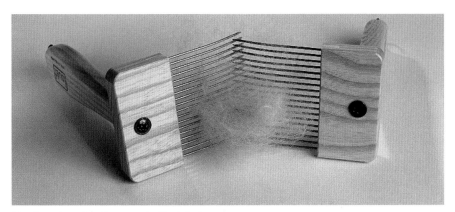

Hand wool combs. Photograph: Nigel Swift

Raised woollen cloth and twill woven from tightly twisted worsted thread

Filpucci of Italy. A modern spinning factory producing high fashion yarn

During the 18th and 19th centuries, Yorkshire gradually overtook East Anglia in the production of worsted, because the faster flowing rivers were able to drive the milling machinery, and there was not quite as much bitter opposition to mechanisation as there was in East Anglia. Yorkshire became divided into areas famous either for the carding of short crimpy fleece into woollen yarn, or the combing of the long lustrous fleece spun into worsted yarn.

Differences in wool and worsted yarn

WOOL	WORSTED
Spun from carded, rolled wool	Spun from parallel, combed fibres
Fibres are not parallel but cross in all directions	Short fibres combed out long, lustrous fibres left
Elastic	Inelastic
Bulky	Smooth
Garments made from wool are warmer because woolly surface traps air	Makes sleek, shiny textiles with clear woven detail
Finishing can 'bed-in' threads and give soft, spongy texture	Finishing can give smooth lustrous surface
Long draw method suitable for hand spinning	Short -draw method suitable for hand spinning

Woollen and worsted yarns are woven into completely different types of wool cloth.

Spinning today

The two kinds of spinning machines developed in the 19th century, the ring frame and the mule frame continued to be used for staple fibre spinning through the 20th century, and are still the basic design of modern spinning machines today. Modern machines have been designed that run at increasingly high speeds with hundreds of spindles whirling at thousands of revolutions per minute, and the safety demands of modern factories and workplaces mean that the spindles are completely enclosed and relatively quiet.

Published by

ASSOCIATION of GUILDS of WEAVERS, SPINNERS & DYERS

Reg. Charity No. 289590

All good wishes
for Christmas
and
the New Year

Margaret

Susan Foulkes, *Online Guild*
"Warp faced double weave,
1689 silk warp ends inspired by Ramesses girdle".

Printed by Budget Printing Services, Lincoln - 01522 546488

TWO

The Materials

A great many fibres can be used for spinning into yarn. They fall into two main categories; natural fibres which have been used for textiles for at least 4,000 years, and synthetic fibres which began when artificial silk was invented in the 1890s.

Properties of textile fibres

All textile fibres have a characteristic structure, whether they are natural or manufactured. They are all made up of long chain molecules called polymers which gives them a high ratio of length to thickness, strength and flexibility that makes them suitable for spinning into yarn. A textile cannot be stronger, warmer or more elastic than the fibre that goes into the yarn it is made from.

Polymers – long chain molecules

Fibre fineness

Fibre fineness is measured in tex or microns, and determines which use the fibre will be put to, for example, yarn for clothing uses fibres of 50/-/100 tex. Fine fibres will provide a more reflective surface and are more flexible, but are more easily damaged.

Fibre length

The average length of the staple fibres in a fleece can be measured and classified. The 'staple length' refers to the limit of fibre growth. The staple length or fibre growth will vary in wool, cotton, etc. according to climatic and geographic conditions, and generally the longer the length the higher the fibre is graded. Filament fibres are of unlimited length.

Lustre

Fibres with a circular cross-section and uniform diameter are the most lustrous. Finer fibres provide more reflective surfaces, for example, silk and viscose fibres.

Fine, lustrous silk fibres with reflective surfaces. Photograph: Nigel Swift

Tensile strength

This means the amount a fibre stretches before it breaks. Special apparatus is used in textile laboratories to measure the percentage a fibre will extend before it breaks. Taking the fibres fineness into consideration as well as its tensile strength is expressed as G/tex.

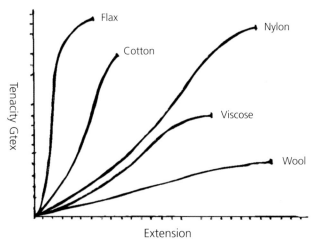

G/tex and extendibility. Flax is strong but not elastic, wool is not as strong but is very elastic and will extend considerably more before breaking

Extendibility and elastic recovery

Yarn and clothing can only stretch and recover their shape if the fibres they are made from have extendibility and elastic recovery, which in turn depends on how straight and parallel the chain molecules inside the fibre are. Wool is moderately strong but very extendable, cotton is strong but not very extendable, nylon is strong and extendable.

Moisture absorption

Absorbent fibres are much more comfortable to wear, as well as being more shrink resistant.

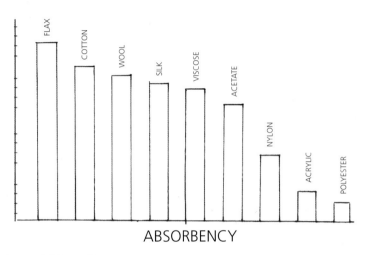

ABSORBENCY

Absorbency of different fibres

Moisture is absorbed from the atmosphere and from the skin. Even when doing nothing the body transpires 1 pint of water per day, and if temperature or physical activity increases this rises.

Resistance to stress and strain

How much the fibre can be bent and twisted without breaking is measured by friction tests in the textile industry.

Conductivity

Heat insulation of clothing relates to the conductivity of the fibres. Wool fibres are poor conductors so wool feels warm; linen fibres are good conductors therefore linen feels cool.

Static

The less static fibres accumulate the more suitable they are for clothing. Cotton, linen and viscose do not accumulate static, wool very little, polyester and acrylic accumulate the most. Static causes clinging, sparking, crackling and dirt attraction in garments.

Natural fibres

Natural fibres are obtained from animals (proteinic) and vegetables (cellulosic), they are all staple fibres with the exception of filament silk.

Proteinic fibres

Animal fibres are obtained from living things, so they vary between young and old and from one animal to another.

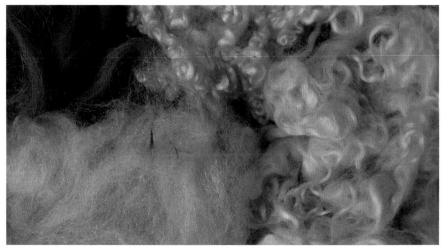

Wool fibres. Clockwise from top left, combed Shetland moorit, Norwegian Spielsau, White Wensleydale, Dorset Horn

Wool

Wool is the ideal fibre; it is warm, elastic, water and fire resistant and has a natural crimp which is partly retained in the spun thread, trapping air and increasing warmth. Wool yarn makes hard-wearing fabric which does not crease but will hold deliberately pressed creases. The porous surface of the fibres mean that they will absorb dye well. The Woolmark is the international symbol used on 100 per cent wool clothing.

Selective sheep breeding has for centuries improved fleece and eradicated the hard outer hairs or 'kemp' and there are now innumerable breeds of sheep whose wool varies in colour, crimp, lustre and length, and which are suitable

Merino, Cheviot and Wensleydale wool dyed with Madder. Photograph: Nigel Swift

for different sorts of textiles. Wool is categorised by the staple length, a measurement from root to tip of between 2 and 15 inches (5-25 cm). Commercially, wool yarn is measured by the Bradford wool count, which is the number of 560 yard lengths to weigh 1 pound (of the finest thread possible). A count of 50-60 is the finest down, whereas 28 denotes a heavy breed.

Fleece when shorn is oily, this 'suint' comes from a gland at the base of each fibre and it weatherproofs the fleece. Suint is a chemically complex substance, containing proteins and ammonia, and can act as a natural detergent in cool water. If cleaning is kept below 40 degrees, some natural grease will remain to keep the wool weather and waterproof.

Silk

Silk is the fibre of luxury and wealth, and is used to make the most valuable textiles in the world. Despite its softness and fluidity, weight for weight, it is stronger than a steel rod, rot-proof and antibacterial.

Silk fibre is spun by the larvae of the silk moth. There are 80 species of silk-worm, of which 13 species are cultivated; *Bombyx mori* is by far the most historically and commercially important. The eggs of the silk moth (35,000 of which weigh 25 g) are hatched in an incubator and the worms of this species fed on mulberry leaves. The silk worm changes skin four times as it grows, before starting to spin the web for its cocoon. Pale yellow gum is secreted from orifices in the head of the larvae and it will extrude a filament more than one mile long, and wrap it around itself. In the right climatic conditions it will spend 15 to 22 days in the cocoon before hatching as a moth. 100 g of cocoons will yield 20-25 g raw silk, which is turned into two sorts of silk thread; filament silk sometimes called grège or frisson silk, and bassinet or 'waste' silk.

The cocoons are put into water at 80 degrees and stewed for 20 minutes, causing the sericin gum to soften. They are then agitated, causing the ends of the filaments to detach and float free from the cocoons. Between 3 and 20 gossamer threads are drawn off the cocoon onto a square reel together, at this stage the silk will stick to itself, so must be separated. The light filament threads are still too fine to use and are ready for 'throwing' or twisting together. After throwing, the thread can be used for silk floss and is still too

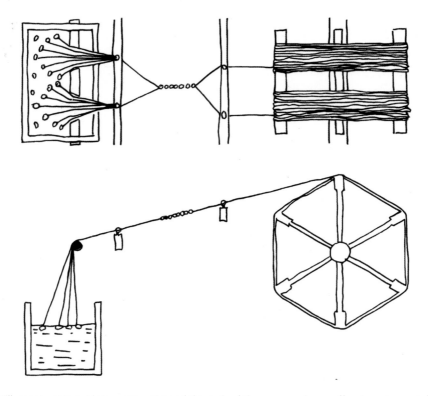

Silk cocoons are put into warm water and the ends of the cocoons drawn off onto a square reel

delicate to be woven. After throwing and plying, the silk filaments are ready to make the sheerest silk chiffons, and high twist (organzine) warp thread.

Bassinet (literally basin waste from the cocoons left in the basin) is spun from the 40 percent of silk left on the cocoons after the filament has been reeled off. This silk is boiled to remove sericin, combed and carded and spun into long staple yarn, of different grades depending on the length of the filaments. Burette and noil silk spun from damaged cocoons are the highest denier and therefore lowest grade silk threads. Higher grade Bassinet is used for Japp, Habotai and silk velvets.

Tussah silk is produced by the caterpillars of various Antharaea moths (*Antherea pernyi, paphia,* and *yanamai*). The silk filament has an irregular surface and a light golden colour. It was traditionally uncultivated and therefore the cocoon was broken at the emergence of the moth, giving shorter filament lengths.

Silk throwing by hand on a wheel

Shantung and Tussore are woven from wild silk, which was first seen in Europe at the Paris exhibition of 1888. The wild silk cocoons are degummed, opened under water and stretched on a frame to make Mwata squares, or on a porcelain dome to make a 'cap'.

Mohair

The lustrous fibre of the Angora goat, mohair means 'best fleece' in Arabic. Native to Anchora in Turkey but now successfully bred in the USA and Australia the Angora can be sheared twice per year, giving a fleece of strong fibres with few surface scales up to 30 cm long. The best fleeces are from animals aged 6 months to 4 years. The absence of crimp means that the fibres do not felt and the

Tussah silk fibres. Photograph: Nigel Swift

best kid mohair can be less than 30 microns think (human hair averages 60 microns thick).

Cashmere

Cashmere is the undercoat of the Tibetan down goat, known as Pashim goat in Persia, and used for shawls called Pashmina. The outer hairs are coarse, but the undercoat, once de-haired, is soft and fine with fibres up to 9 cm long. Each animal only produces 100 g of these warm and luxurious fibres each year. As pasturage and climate affect the fineness of the fleece, attempts to breed Cashmere goats in Europe and Australia have been unsuccessful.

Alpaca and llama

Both alpaca and llama produce a fleece of fine soft hair which can be up to 60 cm long. The llamas and alpacas of the Andes mountains of Peru have been used for clothing from pre Inca times. They produce a range of lustrous coloured fleeces from cream to jet black.

Vicuna

The vicuna from the high Andes produces the finest and most expensive hair in the world. It was reserved for Inca royalty in pre-Hispanic times, and spun to incomprehensible fineness. Supposedly the vicuna suit was the ultimate status symbol of New York mafia bosses and the film stars of the 1930s. The vicuna is now a protected species.

Camel

The Bactrian camel produces a two layer coat; the outer hairs are used for the strong warps of rugs and kelims, and the soft inner hairs can be used for garments such as coats and dressing gowns. Mostly collected after moulting,

Angora goat. Vauxhall City Farm

Fine cashmere fibres and yarn

camel fleeces have a natural range of black and brown to caramel colour and are impossible to bleach.

Angora rabbit

The soft fine fibres of the Angora rabbit from Turkey and the deserts of Jordan can be white or a range of colours. Often blended and used in knitted garments, the fibres can be 10 cm long and are excellent heat conductors, keeping the rabbits cool by day and warm in the cold desert night.

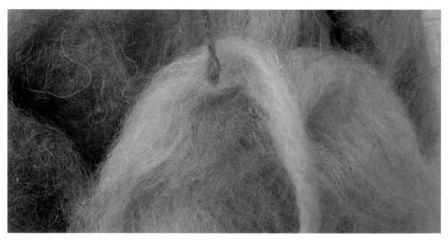

Camel hair and camel under down

Cellulosic fibres

Flax

The flax plant *Linium usitatissimum* is an annual, producing five petalled blue or white flowers. The *Silene lineold* variety grows in most cool parts of Europe, and the historic flax of Egypt and Mesopotamia is the *Silene linicola*.

Flax produces *bast* or *phloem* fibres, which means that the fibres occur in bundles in the stem of the plant. Each bundle has 30-40 fibres and each stem has up to 30 fibre bundles. When the fibres are converted into yarn it is known as linen. Because the fibres conduct water from the root to the top of the stem and allow it to evaporate, linen clothing is absorbent and the best fibre to wear in hot climates. Linen is strong, hard wearing, lustrous and, although easily creased, it irons well and dyes well.

Traditionally sown on Good Friday (closely sown to produce tall slender plants), flax was ready to harvest in August when it was rotted down in dew tanks for 20-30 days (or slowly moving water for 10-14 days). Flax is now rotted with chemicals, which is quicker but more expensive. After rotting or 'retting', the fibres freed from the decomposition go through several beating and combing processes called 'scutching' and 'hackling' to completely break down the woody substance and prepare the super strong fibres for spinning.

As the flax dries, it naturally rotates in an 'S' or clockwise direction and has therefore been traditionally spun in that direction. Strong, but not very extendable or elastic, linen has high fabric shrinkage, which can be reduced by finishing processes. The famous dazzling whiteness of bleached linen has prevented it being blended with other fibres, as linen used to be bleached after weaving, but recent developments have enabled the fibres to be bleached before spinning.

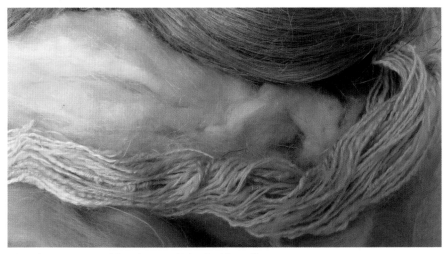

Fibres from a variety of flax plants and bleached linen fibres

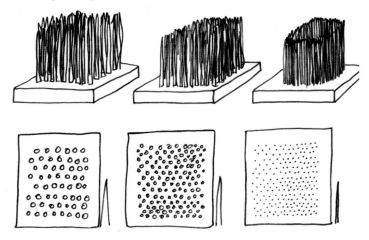

Flax combs showing teeth of decreasing gauge

Cotton

A member of the Mallow family, the cotton plant in the Americas (*Gossypium peruvianum, hirsutum* and *barbadense*) and in Asia (*Gossypium herbaceum* and *arboreum*) produces cotton as hairs round the seeds, 4000 to each seed. The cotton fibre is a thin tube that flattens into a flat ribbon when it is ripe. When the fruit opens it resembles a snowball of fibres, which have to be disentangled from the seeds by 'ginning'. Used in India and Assyria from 2000 BC and brought to the West by Arabic traders, the word probably comes from the Arabic *qutin*.

Cotton fibres are good conductors of heat, allowing water to evaporate and do not build up static electricity so are very comfortable to wear. Although not very elastic, they are very strong, 25 percent stronger wet than dry so can be twisted tightly for warp thread. Cotton fibres react well to laundering at

Bleached and unbleached linen yarn, cotton fibres and fine cotton thread

high temperatures, dyeing with indigo and synthetic dyes and mercerising to give sheen; they are also flame resistant. West Indian and Egyptian cotton have the longest staple length of 40-65 mm, Indian cotton is less than 20 mm. Cotton has innumerable uses in clothing: denim, voile, winceyette, towelling and lace for all sorts of garments from T-shirts to underwear.

Hemp

Hemp (*Cannabis sativa*) from the nettle family native to Asia, is a cool weather annual producing a strong bast fibre. Hemp is retted in ditches and pulled through combs, and it can be dampened and spun in the same way as flax, although it will not produce as fine a yarn. It is usually bleached and woven into carpet backing and rug warps. From the middle ages onwards sails, canvas and ropes were made by guilds of hemp spinners and weavers. Recently a lighter fibre called 'Manilla hemp' (from the banana family) has been used for clothing, often blended with other fibres.

Jute

Jute (*Corchorus capsularis*) grows to 2 metres and has to be cut by hand. It has the same retting and drying process as flax, but retting takes 30 days. Jute cannot be bleached and has been used for carpets, sacks and upholstery throughout textile history. In Scotland, Dundee was the traditional centre of the jute industry. In the 17th century Flemish weavers developed mixtures of flax wool and jute for tapestries and upholstery.

Sisal

Sisal (*Agave sisalana*) is a heavy cellulosic fibre obtained from the plant's leaves. The fibres are up to 10 cm long and are dried out and dampened before spinning. The creamy white fibres are used for household items and upholstery.

Nettle

The common nettle (*Urtica diorica*) is being developed in as an environmentally compatible textile fibre. Ramie nettle (*Boehmena nivea*) is an environmentally friendly fibre, the nettle is a fast growing and high yielding plant and does not deplete the soil of nutrients. Ramie grows in hot rainy climates, the woody bast fibres are obtained after 'boiling off' the plants and de-gumming the long fibres, which are strong but have no elasticity. When dried the fibres are brilliant white, very long and strong. Called 'China grass' or 'China linen', ramie has mostly been exported for use in upholstery, but is now increasingly being used for clothing (sometimes blended with viscose) as it has qualities of moisture absorption and conductivity similar to those of linen.

Hemp, jute and sisal fibres

Bamboo

Bamboo fibres have been the phenomenon of the first years of the 21st century. Biodegradable and eco-friendly the fibre is produced in the Hunan province of China. Cool and comfortable, Bamboo fibre has been validated as antibacterial, even after extensive washing, and it is cool and comfortable to wear in hot weather.

Paper

Although not a fibre, paper yarn is twisted from ribbons of paper made from three plants (*kozo*, *mitsumala* and *gampi*) using traditional Japanese methods. Paper yarn is soft and warm with a delicate sheen and allows air to pass through it. It is easy and cheap to process but rather fragile and must be washed with care.

Clockwise from top left, bamboo, soya, ramie nettle and nettle fibres

Manufactured fibres

Manufactured fibres now account for 70 per cent of the fibres in the world made into yarn, although it was not until the 1940s that they became significant in the textile industry. From the creation of artificial silk by Count Chardonnet in the 1890s, and Joseph Swan's nitro-cellulose embroidery threads for his daughters patented in 1883, scientific research has produced a new range of fibres of high functionality produced for specific uses. Manufactured fibres fall into two categories: regenerated fibres and synthetic fibres.

Regenerated fibres

Rayon and viscose

Rayon and viscose are generic names. They are derived from pine or spruce chips, steamed for 15 hours, then turned into a treacly mixture by the addition of caustic soda and forced through spinnerets. Used for linings, the filament is smooth and bright and can be used for satins and brocades. Trade names include Tricel, Tencel and Lenzel.

Acetate and Tri-acetate

Invented in 1921, acetate and tri-acetate are derived from cotton bits and 'linters' which are dissolved in acid and made into spinning 'treacle'. Acetate has good draping qualities, is very extendable and does not hold static. Trade names include Dricel Lyocel (Courtaulds) and American Lustron.

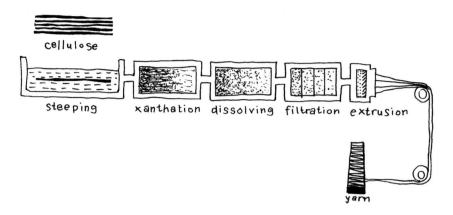

cellulose

steeping   xanthation   dissolving   filtration   extrusion

yarn

Viscose spinning; cellulose is rendered down into the viscose solution and extruded through spinnarets

Synthetic fibres

Synthetic fibres are formed entirely by chemical synthesis. The polymers or chains from which they are made are artificially produced. After the creation of nylon by DuPont in 1935 two generations of synthetic fibres have subsequently been developed; the polyester and polyacrylic fibres of the 1950s, which were chemist's copies of natural fibres, and from the 1980s synthetic polymers of high strength and flexibility.

Nylon

The nylon polymer is made from benzene (from coal) plus hydrogen (from water). Chips of the polymer are then melted and the filament is extruded by the spinnerets. When it is stretched and drawn it becomes elastic. Nylon has no thermal qualities unless the surface is brushed up, but it is hydrophobic (repels water) so dries well and does not burn. Nylon can be crimped by heat processing to add extendibility. Trade names include Tactel.

Polyester

Polyester is made from petroleum, ethylene glycol and trephthalic acid. It has poor elasticity but resists creasing and water, so drips dry. It builds up static and does not dye easily. Polyester can be textured and crimped. Trade names include Crimplene (Courtaulds), Trevira (Monsanto), Dacron (Du Pont) and Terylene (ICI).

Acrylic

The acrylic filament is made from acriylonitrile, a liquid derived from oil, synthesised with carbon, nitrogen and hydrogen before being melt-spun. The extruded filament is usually bulked by steaming and cut up for staple spinning. Because of its soft, warm handle, acrylic is used for clothing, often as a knitting wool substitute. Resilient and shape retentive, it is not, however,

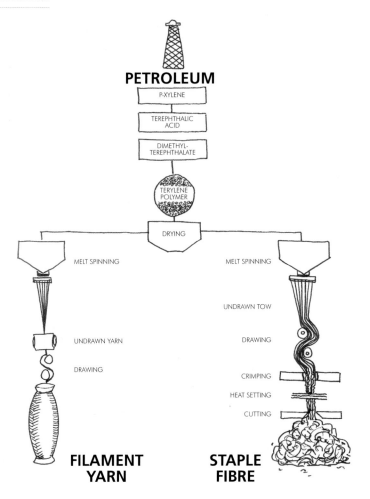

PETROLEUM

P-XYLENE

TEREPHTHALIC ACID

DIMETHYL-TEREPHTHALATE

TERYLENE POLYMER

DRYING

MELT SPINNING

MELT SPINNING

UNDRAWN TOW

UNDRAWN YARN

DRAWING

DRAWING

CRIMPING

HEAT SETTING

CUTTING

FILAMENT YARN

STAPLE FIBRE

Extrusion of synthetic filament fibre

thermal or absorptive and can build up static. The soft lustrous properties make it suitable for the pile of fake fur fabrics. Trade names include Acrilan (Monsanto), Courtelle (Courtaulds), Dralon and Orlon.

Elastane
Elastane is a polyurethane fibre used for Lycra (Du Pont).

Aramid fibres
Aramid fibres are the strongest fibres in the world and were developed for the space shuttle. They can be cut lengthways into gossamer fine microfibres. They have the trade name Kevlar.

Lycra stretch top

Bi-Component fibres

Bi-Component fibres are nanofibres that can be split into 16 sections, often with a nylon core and polyester coating. Each fibre is less than 0.18 of a denier and the filaments have a silk-like touch.

The identity of a fibre

The identity of a fibre can be difficult to ascertain, especially if a yarn is made of mixes and blends of fibres. To identify a fibre in cloth, separate the warp and weft or unravel knitted fabric. Untwist the threads and examine them through a magnifying glass when all the threads are untwisted and parallel.

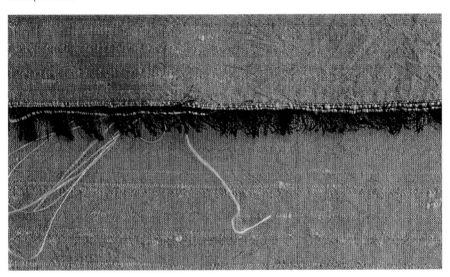

Removing weft filament to analyse fibres

THREE

Yarn spinning mechanisms

A handful of fibres can be simply teased out and twisted into a yarn without the use of any tools. Finger spinning has been used since earliest times to make and ply threads, and is still used today in some parts of the world to twist loose rovings for rug and blanket weaving. Whatever apparatus is used, yarn is produced by twizzling threads as they are drawn out or 'drafted' and by twisting the extended thread round the stick to store it.

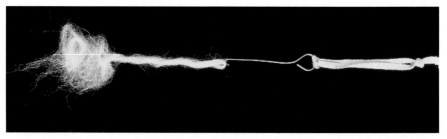

Twisting threads as they are drawn out by a hook. Photograph: Nigel Swift

The spindle

The spinning spindle is a small tool turned by hand for twisting fibre and has not changed for 20,000 years. Spindle spinning produced all the staple thread in the world until the wheel began to be used, and even then the first European wheels could not produce a strong thread, so warp thread had to be spun by spindle until the appearance of the 'Saxony' or treadle wheel in the 16th century. The wheel never really replaced the spindle, which is still used today in parts of the world and was common in Europe until the 1930s. In the USA today a new generation of high-tech supported spindles on ball bearings are used by textile makers as an efficient, portable way of making specialised yarn without stress or noise.

Not all spindles are drop-spindles, that is suspended by the yarn and revolving in mid air. Traditionally some North American Indians spun yarn using large-whorled spindles supported on the ground. The Indian cotton spindle was supported in a bowl to prevent the fine thread breaking from the weight of the spindle, and in some areas spindles are rolled down the thigh as the fibres are drawn out. Balkan spinners who use the suspended spindle have two whorls on the shaft to keep the spindle balanced as they walk about herding sheep.

The experienced spinner will be able to to twist the shaft (which is the axis), so that the energy of the spin and weight of the whorl will keep the spindle spinning steadily without wobbling throughout the drafting process.

Using the drop spindle, drawing out thread

The yarn has to be made in two stages on a spindle as spinning has to stop to wind the yarn around the shaft. The weight of the whorl is very important; as it keeps the spindle turning. Larger whorls give a long, slow twist to make heavier yarns; whorls of small diameter give a tighter twist to lighter, finer yarns. Drop spindles without whorls can put in even more twist because there is no friction to slow them down.

Spindle spinning has provided the yarn for centuries of textiles, and is also a contemporary textile skill requiring speed and co-ordination between eye and hand.

The spinning wheel

The first time a wheel was used to wind yarn was in China, where the silk from cocoons was wound onto a square wheel. From illustrations, it appears that a hand turned wheel was also used to twist silk threads. The small cotton-spinning wheel or *Charka* has also been in use for hundreds of years in India. Both of these mechanisms were turned by a handle and were used on a small scale. By the 14th century, early hand-turned wheels had spread to various parts of Europe in different forms. A spinning wheel is first mentioned by the Abbot Usher at Speier in 1320, and first illustrated in Flemish manuscripts of 1338, becoming the 'Great Muckle Wheel' in Scotland. The English wheel was often turned by a rod or 'wheel finger'. As the spinner walked a few steps from the wheel holding the wool straight out from the tip of the spindle, the long stretched roving would twist as the wheel turned, then a reverse flick would unwind the yarn from the spindle tip and,

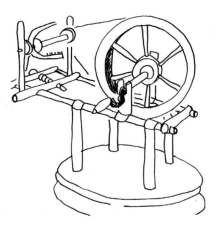

The hand-operated flyer wheel

The U-flyer mechanism allowing simultaneous twisting and winding on. Photograph: Nigel Swift

holding it horizontally to the spindle, another turn would wind it on.

The large early wheel gave a continual and even rotation, but spinning was still intermittent as twisting and winding on were still separate operations.

The U-Flyer wheel led to the great innovation of simultaneous twisting and winding on that is still used in mechanised spinning and twisting. A 'U' shaped arm was fitted on to the spindle and this distributed yarn around a bobbin that was also fitted on to the spindle. Both bobbin and flyer rotate, but the flyer rotates faster, wrapping yarn on to the slower moving bobbin. As the bobbin fills up, its diameter increases and the yarn receives less twist before being wound on to it, so there must be a method of adjusting the rate of winding on by changing the relative speeds of wheel and bobbin.

The inventor of the flyer wheel is unknown. A hand operated flyer mechanism, the 'Travelling Flyer' is shown in the drawings of Leonardo Da Vinci in 1490 in his Codex Atlantica; he might have been influenced by the Italian silk throwing mechanism. His system of driving the bobbin and flyer by larger and smaller cogs, was actually built when the textile industry became mechanised three centuries later.

The earliest actual illustration of a wishbone-shaped flyer is from Brunswick in Germany, dated 1520. The wheel design which became known as the Saxony wheel, and is illustrated in the Hausbuch of Waldburg in the 1520s, was further improved by the addition of the foot treadle early in the 17th century, again the origins of this innovation are undiscovered.

The foot treadle left both the spinners hands free and meant a continuous flow of fibres of consistent quality could be twisted and wound on until the bobbin was full. The spinner only had to prepare the roving to the thickness required, draft in the fibres and adjust the tension.

Wheels of all shapes and designs were used throughout Europe and America for the next 400 years. Different wheels evolved for different fibres, double treadles and distaffs were added and even extra wheels, but all spun just a single thread. Most pre-industrial spinners specialised in one type of yarn, usually regional specialities for local fabrics, and became experts at intuitively judging the turns of the wheel and the twists per inch of each tension, so that no written instructions were needed for the construction of the yarns.

Early 17th century foot treadle U-flyer wheel

Machine spinning

The principles of the spinning wheel influenced the first generation of spinning machines. In fact, the mule which is still used for the specialised spinning of cashmere in India and China, is based on the spindle spinner. The carriage draws out the thread, slipping it off the end of the spindle to insert twist.

The flyer mechanism used in ring spinning was invented in the USA in 1929 it is very much like a vertical version of a spinning wheel. It is still used for mohair and soft wool weft yarns. Thread is thrown around the bobbin by a 'U' shaped flyer, reversing direction for plying and multi-plying. The Balloon spinner works in the same way, but the yarn balloons out.

Rotor spinning is a system of rotating cups into which clumps of fibre are sucked. As the fibres are twisted they are pulled out of the bottom of the cup, rather like pulling clothes out of a washing machine or dryer while it is still rotating.

More recent innovations do not owe so much to the technology of the wheel. Break or open-end spinning, invented in the 1980s, relies on the fibres rather than the package or spindle being twisted. Fibres forming a continuous

Modern spinning machines; Filpucci, Italy

trail of yarn are rotated by a vortex of air as they pass though a tube. This can be done at high speed, up to 80,000 revs per minute, using less power and fewer operatives, and with less waste.

Further developments in the 1990s developed this method further. Jet spinning and air spinning can automatically rejoin breakages and replace full bobbins. All mechanical spinning methods use a formula to pre-set the twist angle from the axis and maintain the count. The machine is reset to change the character of the yarn.

Spinning machines spin staple fibres, so manufactured filaments and silk have to be chopped into staples to be spun or blended at the drawing out stage. The preparation of fibres for mechanical spinning involves combing on a cylinder, and then drawing out, in which the carded twist-less filaments are passed from slowly rotating back rollers to higher speed front rollers. As the fibres pass through the drafting zone the roving becomes longer and thinner.

FOUR

Handspinning Techniques

Handspinning is thought of as a primitive folk art, but it has been a necessity in producing textiles of great sophistication and beauty throughout history. All textiles before the late 18th century and very many of the 19th century were made from yarn spun individually by hand. An understanding of the construction of yarn is essential to making a spun thread that will have the right twist and balance and the right fibres for suitable handle. It also enables us to appreciate the astonishing skill of the spinners who produced the intricate wools of the North and the fine linens of Pharaonic Egypt, woven at 100 ends per centimetre.

The technicalities of yarn construction, vital for a hand spinner constructing their own yarns, are also useful and necessary for any textile practitioner. The effect on fabric of the twist and ply of the yarn, the opposing energies of multiple plies and the decorative surfaces that are the result of specialised yarns, can be assessed by trying out some of the construction techniques. Hand spinning enables any textile designer to understand the behaviour of yarn in or on a fabric.

Why handspin yarn?

Although machine made yarn is produced increasingly rapidly and is constantly innovative, making yarn by hand can provide the designer with many benefits. Specialised handspun yarns can dictate the design of the fabric, creating exactly the colours and textures required by the knitter, weaver or embroiderer. The handspinner can change the scale, colour and fibre at will to create complementary ranges of yarn in any amounts from a few grammes to multiple skeins, producing a quantity that is exactly right for a particular textile. Using equipment of ancient and unchanging design, handspinning can be a genuinely creative design facility rather than just a laborious pre-industrial process.

Designing handspun yarns

When working on a range of yarn designs, small amounts of each construction can be made from pre-dyed fibres, using the coloured fibres like a paint palette.

Complementary range of hand spun yarns. Photograph: Graham Murrell

The twist angle, plying patterns and distance of repeating features can be varied in small sample skeins by winding singles on to hexagonal pencils or squared pieces of wood, which will give tension and provide enough weight to ply these small sample lengths.

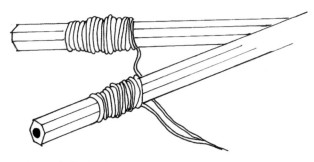

Wind singles on to pencils for tension

In designing and spinning yarn, three major elements have to be taken into account: fibre, twist and draft. The character of all yarn is dependent on combinations of these, and altering them will affect the handle and strength of the yarn and therefore the fabric made from it.

Fibre

The character of the fibre; its length, lustre, softness and crimp, will affect all the other considerations in designing yarn. The great variety of fibres available can be exploited for their design potential and can be blended and mixed or plied together as singles to give mixtures of lustre and length or crimp and softness. The same construction and twist will behave completely differently in different fibres, even in different varieties of the same fibre.

Mixed silk and mohair plied with silk. Yarn hand-spun by Ann Wasiliewska.
Photograph: Graham Murrell

Twist

Calculating the degree of twist from the twist angle and the twists per centimetre gives an understanding of the underlying structure of the yarn. The advantage for a spinner is that matching batches of yarn, repeating designs and continuity throughout the yarn can be more accurate if the twist can be checked and maintained.

S/Z

The convention of spinning singles in Z direction and plying in S direction possibly evolved from early linen spinning, as flax fibres have a natural curl to the left (S). However, wool has a vector-less crimp and some fibres curl to the right, so historically the convention may be to do with right-handedness.

Twist angle

Consistent twist angle can be achieved in hand spun yarn by regularly checking it against a protractor. The usual singles direction (Z) spin at the right side and the plying angles twist at the left side. This affects the hardness or softness and draping quality of the fabric the yarn is used to make, and if the twist angle can be maintained, the yarn will have the same handle and strength throughout.

Twists per centimetre

To understand and control the number of twists per centimetre it is necessary to know the 'twist ratio' of the spinning wheel. Whereas a spindle spinner drafts the fibres straight on to the moving spindle, a wheel has its own ratio of the number of twists of the fibres caused by each turn of the wheel. Each type of wheel has its own ratio. In the most commonly used Saxony wheel there will be around seven revolutions of the flyer around the bobbin each time the wheel is rotated by the treadle. Therefore, one treadle puts in seven twists to the drafted fibres.A fast wheel with a high ratio will insert more twist for each treadle before the draft is drawn in. It is easier to spin fine yarn on a fast wheel and bulky low twist on a slow or low ratio wheel. Historically, the ratio of a spinning wheel related to the one fibre it was used to spin. For example, a 17th century worsted wheel has a high ratio of 10:1 (10 revolutions of the flyer to 1 treadle) whereas wheels used to spin silk floss for surface embroidery, reflective and lustrous without strength, would have a low ratio of 4:1 (giving 1 or 2 twists per centimetre). Wheels with a high ratio (used for household linens) could not have been used for soft knitting wools.

The twist ratio means the relationship between one turn of the wheel and the number of times the flyer wraps thread round the bobbin. Photograph: Nigel Swift

Plying twist

Singles have an 'active' twist which has to be counteracted by the balancing ply, the most visible twist in a plied yarn.

The plying twist partly reduces the spinning twist by rotating the singles anticlockwise. It is calculated that 2 fold plying will counter-twist spin by 1/4, so a single of four twists per centimetre would be reduced to three twists per centimetre by a ply of three twists per centimetre. Balance is achieved when

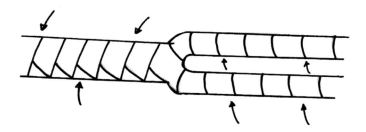

Singles twist balanced by ply

the two singles are turned enough to counteract each other, and it can be seen when a single is allowed to double back on itself until the two spirals interlock.

Draft

The draft controls the amount of un-spun fibre drawn or stretched to be fed into the wheel or on to the spindle; the fibres are stretched into the twisting zone controlling the thickness of the yarn. In hand spinning the drafting zone is between the two hands. Twist angle is determined by the thickness of the thread; the finer the thread the more acute the twist angle for the same strength, so draft, twist angle and twists per centimetre are all interdependent.

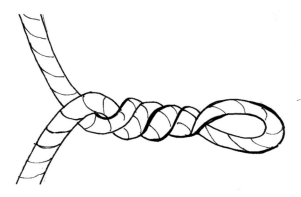

Single doubling back on itself

Using the spindle

Using a drop spindle or a supported spindle requires skill, but can be far from slow and clumsy. The process is best understood by experimenting with drawing and twisting fibres from raw fleece. This finger spinning will demonstrate the relationship between the amount of twist and the length of the draw.

Drop spindles twist more quickly from a vertical draw, but the spun thread has to contain enough twist to carry the weight of the spindle. The supported spindle is rolled in the hand or along the thigh, the draw is horizontal, and a softer, thicker yarn is produced. The drafting zone is between the hand and the supply of unspun fibres, and consistency is maintained purely by eye and touch.

Using the spinning wheel

Using the double or single band treadle wheel, a continuous flow of yarn can be produced until the bobbin is full. Twist is imparted to the fibres from the moment they leave the spinner's grip, until they reach the flyer and are wrapped around the bobbin. The speed is always dictated by the treadling rate.

Twist is imparted to the fibres from leaving the spinner's grip to being wrapped around the bobbin. Photograph: Nigel Swift

The thickness of the yarn depends on the thickness of the prepared fibres, taking the drafting process into consideration. The length of each draft should be the same for each treadle, to produce an even, balanced thread, and can be lengthened or the treadles increased to make an extra twisted or extra loose single.

As spinning wheels were traditionally used to produce one particular yarn, the modern treadle wheel has to be used with a slightly different technique for different fibres and effects. By adjusting the drive band on a double band wheel, or tightening and slackening the brake mechanism on a single band wheel, different weights of yarn can be spun. A fine yarn can be dealt with by slackening the bobbin completely then gradually tightening it until just enough 'pull' is exerted to draw in the fine yarn. The bobbin needs to be tightened to pull in a bulky yarn as quickly as possible (using consecutive hooks on the flyer to fill the bobbin evenly).

Use consecutive hooks on the flyer to fill the bobbin evenly. Photograph: Nigel Swift

The twist ratio of the wheel can be controlled by the speed of treadling. In a thread made on a spinning wheel, the inner fibres interchange position with the outer fibres, rather than the yarn building up in layers as it does in most mechanical processes, which equalises the stress and means that the outer fibres do not wear and shed.

Yarn construction and design

Design by preparation

Part of the design process for yarn is the way the fibres are prepared for spinning. Carding, combing and teasing out affect the appearance, weight and amount of air in the yarn, and hand carding will stretch delicate fibres considerably less than mechanised methods. Carding and combing different types of wool will determine the surface texture of the yarn and the sort of cloth it will make.

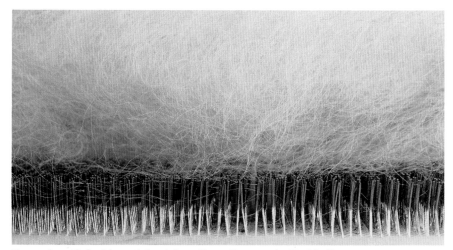

Fine carder with flexible teeth for delicate fibres. Photograph: Nigel Swift

Carding fibres

Fibres with a staple length of less than 10 cm and a high crimp are carded into light, lofty rolags; the fibres at the end of the rolag are drawn off transversely. This traps the most air and produces soft, spongy yarn, spun by long draw for maximum elasticity and lightness.

Light, lofty rolags of mixed colour fibres and the thread spun from them.
Photograph: Nigel Swift

Combing

Combed worsted is prepared by drawing long staple wool through a fixed wool comb using the other comb to pull the long fibres into a sliver, extracting the short ones. The sliver is spun by feeding successive lengths straight

towards the orifice of the spinning wheel (the opening at the end of the spindle). Long fibres can be spun from a fold over the forefinger, very long slivers can be spun from a distaff, enhancing the lustre by keeping the fibres parallel.

Colour carding

If working from pre-dyed fibres, the carding can determine the colour combinations, and mixing or blending fibres can determine the texture.

Nepps

By combining shorter nepps and fragments of contrasting fibres, the spun yarn will be automatically slubbed and textured.

Wool sliver spun from a distaff.
Photograph: Nigel Swift

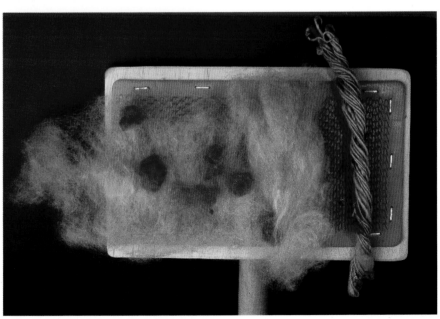

Blobs and nepps carded in. Photograph: Nigel Swift.

Mixed fibre constructions

Blending

By blending complementary fibres, the resulting yarn can have the sheen of silk and the crimp and elasticity of wool. Very small quantities of expensive fibres can be blended with wools to make yarns with a luxurious texture.

A little white carded into coloured fibres makes a range of yarns known in the trade as *chiné*. Two coloured carded blends are known as *jaspé*.

Blending fibres for colour and texture. Photograph: Nigel Swift

Mixed coloured fibres. Photograph: Nigel Swift

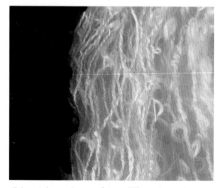

Crimped wool spun from different wool fibres. Photograph: Graham Murrell

Fabric knitted from yarn alternating between silk and merino wool. Photograph: Nigel Swift

Plies of singles

Plies of singles spun from different fibres can crimp and loop if the fibres have different elasticity and recovery from stretch, for example, mohair plied with Merino or Botany wool. Fibres that change intermittently along the length of the single from lustrous to matt, or smooth to raised, knit or weave into fabric of varied texture.

Designs in plying singles

Textured yarns and more complicated multiple thread designs are created at the plying process from sets of singles.

Core thread constructions

These constructions depend on one single being used as the 'core' or anchor, while the other thread or threads are guided round it in a variety of ways to make the design. The degree of twist in the core thread has to allow for untwisting as the other threads are manipulated round it, so plying has to be as near to the orifice as possible.

Gimp

For balanced gimp yarn, a thicker, evenly drawn out single is plied with a thinner smooth single with a little more twist. Gimps are particularly effective in silk.

Thick and thin gimp

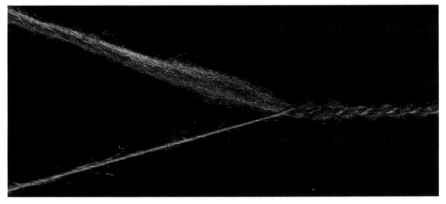

Plying gimp. Photograph: Nigel Swift

Gimps and spirals. Photograph: Graham Murrell

A fine core thread gives scope for the colour and texture in the thicker thread to be highly visible. If the thicker thread is spun from shorter fibres, it will stand out further from the core.

Spiral

An exaggerated gimp, in which the outer thread is several times thicker than the core, is called a spiral. If the singles are separated and held at an angle to each other as they go into the ply, the spirals will be closer together.

Spiral; outer thread thicker than core. Photograph: Nigel Swift

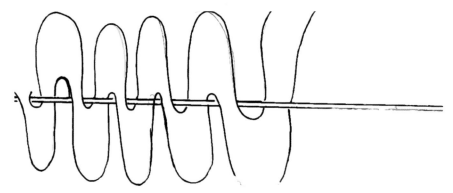

A spiral

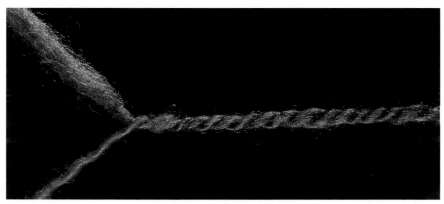

Plying a spiral. Photograph: Nigel Swift

Loop yarn

A high twist smooth fine single forms the core for a thicker high twist looping thread spun from long staple fibre.

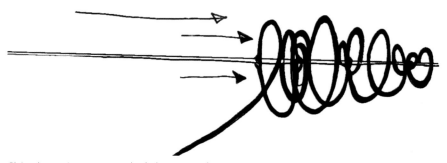

Plying loops. Loop yarn pushed along smooth core

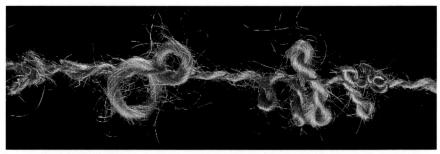

Loops pushed along core thread. Photograph: Nigel Swift

aLooped coloured mohair yarns. Photograph: Graham Murrell

The loop yarn can be pushed along the smooth core, forming patterns of loops that will depend on the fibre used. Mohair will form large regular loops, long wool breeds such as Wensleydale will make varied loops. When spun on a 'jumbo' flyer or spindle, the loops will not catch on the orifice or hooks.

Intermittent loops

Intermittent loops are formed when the looping single is intermittently plied normally with the core. To ply loop yarns, the singles should be placed on either side of the wheel, to prevent tangling.

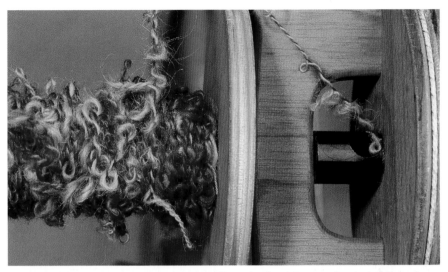

Bulky yarns will not catch on the large orifice of a jumbo flyer. Photograph: Nigel Swift

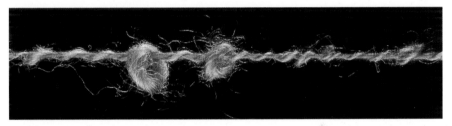

Intermittent loops and plying with core. Photograph: Nigel Swift

Retarded thread constructions

In these constructions one yarn is fed into the ply more quickly than the other. These mixed tension yarns have to be blocked (wetted and dried under tension) or steamed to achieve stability within a woven or knitted fabric. Twist is increased when the single yarn is spun by increasing the number of treadles in each draft

Boucle

The looped and crimped texture of a boucle is the result of a high twist single plied on to a high twist finer single, which is held back or retarded.

The piled ply is allowed to to build up into twists and over-spirals. The nature of the design depends on the fibre; a bright cultivated silk will form crisp swirls, whereas Tussah silk boucle will form softer fuller spirals. If a binder thread is added the yarn is known as ratine, or locked boucle. The over-twist single will be many times longer than the retarded single, which is the case in all yarns of this construction.

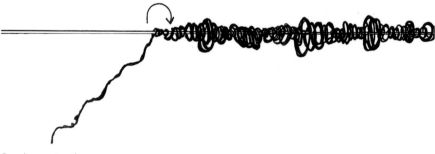

Boucle construction

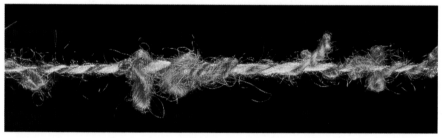

The yellow thread is retarded, the brown thread allowed to twist round it.
Photograph: Nigel Swift

Twists and over-spirals. Photograph: Graham Murrell

Bound boucle. Photograph: Graham Murrell

Snarl or frise

A 'snarl' is formed by a small section of yarn plying back on itself before being plied with the other single thread which is held back as the snarls are formed.

Construction of snarl thread

The spikes are more defined if the doubling yarn is medium fine and has a higher than average twist angle. The bobbin holders have to be kept apart to prevent tangling.

Brown thread pushed into spikes on retarded thread. Photograph: Nigel Swift

Snarl yarn. Photograph: Nigel Swift

Double snarl

In a double snarl the spikes can be made with alternating yarns. Relaxing the thread to form snarls will neutralise the twist which will have to be compensated for in the next section of normal plying. The plies take it in turns to be the retarded thread.

Construction of double snarl

Wraps and knots

Wraps and knots are 2 ply yarns in which one ply is held back and the other wraps around it. Solid wrapping with silk thread can make the surface of the yarn consistently lustrous.

Constructing wraps and knots

Wrapped threads form solid blocks of colour. Photograph: Nigel Swift

If the wrapping ply is pushed into concentrated areas it will form solid blocks of colour. The wrapping ply can be knotted around the retarded yarn in a figure of eight and built up to form knots, held at an angle of 90 degrees from the retarded thread for the wrapping process. In between the knotted or wrapped areas the yarns are plied normally.

Alternating knot

Fine gold knots. Photograph: Graham Murrell

Silk knot yarn. Photograph: Nigel Swift

Designs by adding to the ply

When adding to the ply, the insert will cause an irregular ply distribution which will have to be compensated for in the areas of regular plying.

Inserts

Spots

A 2 ply yarn with coloured blobs is inserted intermittently between the plying threads.

Inserting spots into ply

Longitudinal inserts

For the integrity of the yarn, the blobs must be anchored by extra twist in the plying. The blobs or nepps can be large so that they protrude from the ply either side, or can be put in longitudinally, making an intermittent third ply.

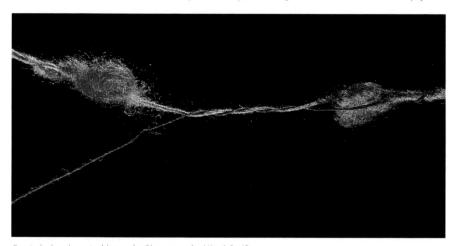

Spots being inserted into ply. Photograph: Nigel Swift

Spots and pieces in ply

Cut pieces

Short pieces can be cut from spun singles and anchored into the ply to give a shaggy or bitty texture.

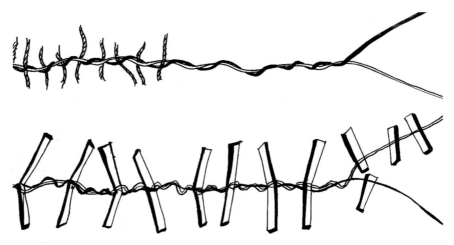

Spun and cut pieces in ply

Fibre pieces in ply. Photograph: Graham Murrell

Spun cut pieces of yarn in ply. Photograph: Graham Murrell

Tufts and spots

Tufts and spots are designs using a mixture of pieces of spun yarn and nepps.

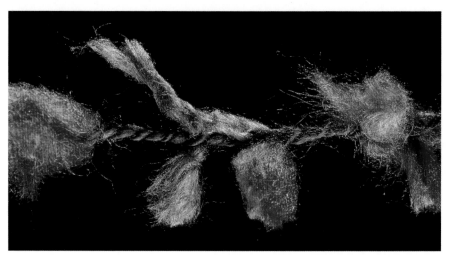

Giant tufts and spots. Photograph: Nigel Swift

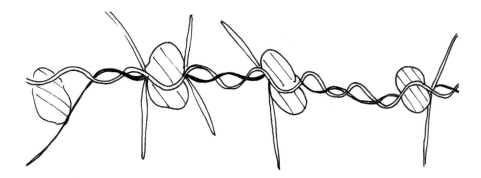

Inserts, spots and tufts

Inserting fibres for fluffy, fuzzy texture

As the two singles are plied, short lengths of sliver are fed into the plying threads. Fibres should be fed in from 'the fold' (gripped between thumb and second finger folded over first finger) for short fuzzy texture. Full length fibres should be used for super-fluff, forming a halo of fibres round the thread.

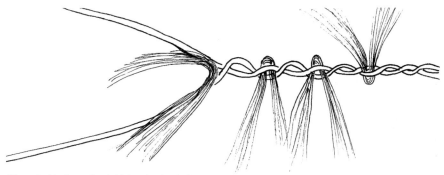

Fibres fed in from the fold for shorter tufts

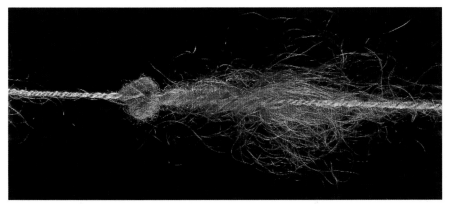

Fluffy yarn with spots. Photograph: Nigel Swift

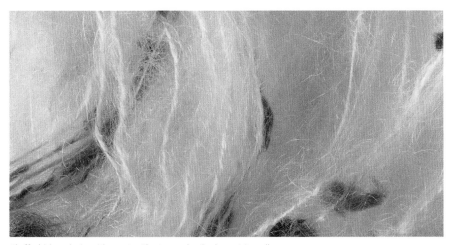

Fluffy kid mohair with spots. Photograph: Graham Murrell

Designs based on slubbing

Designs based on slubbing or deliberately unbalancing the rate of drafting and the twists per centimetre in places along the thread. For this, the fibres need to be held in the middle of the drafting zone, until the fibres behind he slub area have built up enough twist, leaving the untwisted patch to be drawn in.

Short slubs

The length of the slub depends on the staple length of the fibre. Short slubs will be formed by short wool and cottons. However small the slub is, the plying thread should wind right across it to ensure strength in the untwisted area.

Wrap slub in plying thread

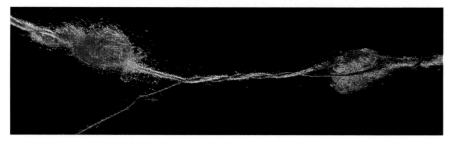

Plying thread winding across slub

Long slubs

Longer fibres will give slubbed thread of an entirely different character, bound by tight twisting of a plying single.

Long slub yarn. Photo Graham Murrell

Double slub

Two yarns are slubbed in different places to create a double slub. This construction is a popular effect in machine spun thread.

Double slub

Long slub and double slub yarns

Multiple plies

Three or more plies can be combined for warmth and strength, or to create intricate designs. The dynamic to untwist increases with the number of plies, so each thread should have extra twist. The amount of air held between the plies increases with the number of threads, giving warmth and adding to waterproofing.

Multiple ply yarns. Photograph: Graham Murrell

Multiple plies can be plied simultaneously, or in pairs or fours. For more than four threads, a system to separate and control the threads is essential.

3 ply construction 'S'/'Z'

Exploiting the tension in twisting and plying this construction is used for many commercial yarns. Two 'Z'-spun singles are 'S'-plied in the usual way, a separate single is 'S'-spun, then the three threads are 'Z'-plied. The construction resembles a plait of criss-crossed threads around a core.

3 ply S/Z construction

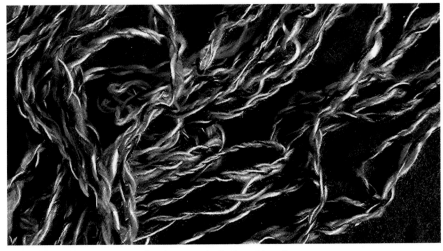

3 ply coloured cashmere and silk. Photograph: Graham Murrell

4-ply constructions

4-ply cable

This is a four component yarn. The 'Z'-spun singles are spun into 'S'-spun 2-ply in pairs. The pairs are then 'Z'-plied, forming an interlocking cable yarn.

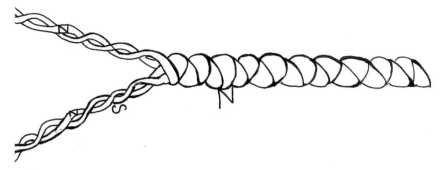

4 ply cable; Z singles S plied then Z plied

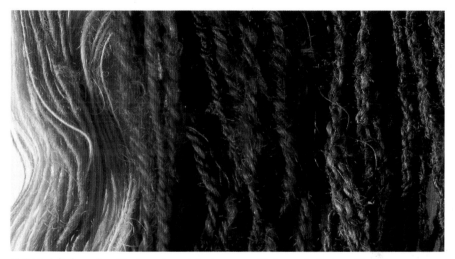

Right to left, four ply cable, two ply and singles yarns.. Photograph: Graham Murrell

4-ply crepe

The 'Z'-spun singles are 'S'-spun into a 4-ply crepe, which will be springy and elastic, but with a smooth surface.

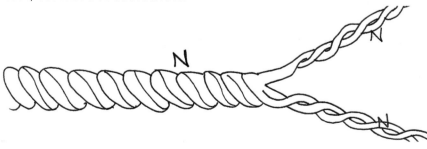

4 ply crepe singles spun simultaneously

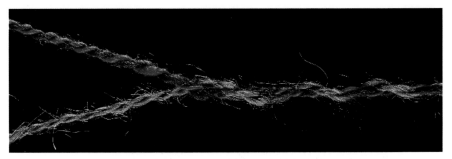

4 ply crepe. Photograph: Nigel Swift

4-ply 'Z'-spin 'Z'-twist cord

This is used for linen warp yarns and multiple plied yarns which have to have immaculately smooth surfaces. Four 'Z'-spun singles are simultaneously plied in the same 'Z' direction. Line flax spun from distaff should be spun wetted with warm water or flax glue (10 g of flax seeds boiled in 1 cup of water). The fibres should be drawn down from distaff and smoothed upwards with damp hands, so that the twist runs up the thread towards he distaff.

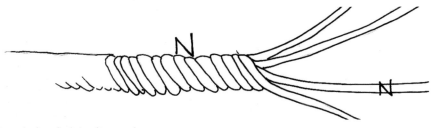

Four singles plied simultaneously

From right: 4 ply simultaneously plied in the same direction, 2 ply, singles from combed wool. Photograph: Graham Murrell

Mock chenille

To create mock chenille, two pairs of singles 'Z'-spun 'S'-plied (with one very fine and one thick) are 'Z'-plied. The thicker threads will bulge and loop out of the cable, giving the texture of uncut chenille yarn.

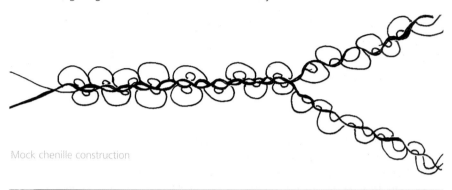

Mock chenille construction

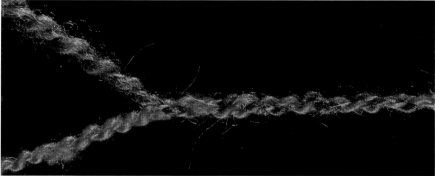

Mock chenille yarn. Photograph: Nigel Swift

Mock chenille. Photograph: Graham Murrell

Navajo ply

This traditional construction means that a single ply can be looped into 3-ply for strength and to give intermittent solid blocks of colour and texture.

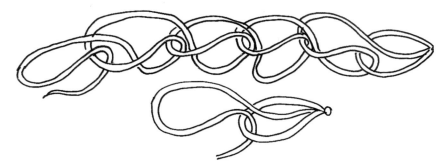

Navajo ply construction

Navajo ply. Solid blocks of colour. Photograph: Graham Murrell

Multiple plys

Multiple plys can be plied simultaneously, or in pairs, or fours. For more than four threads, a system to separate and control the single threads is essential.

For multiple plies the singles yarn should be separated on bobbin racks

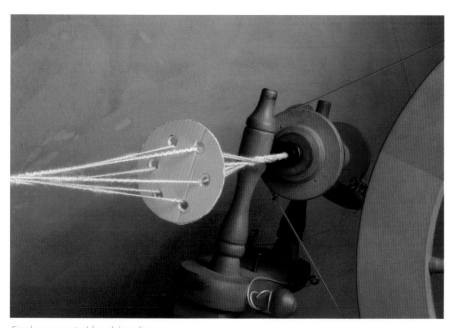

Singles separated by plying disc

Multiple plies. Hand spun by Prue Bramwell Davies, Nicky Hessenberg and Tricia Julian.
Photograph: Ian Hessenberg

Brushed and raised constructions

After spinning and skeining, long hair wools and mohairs can be brushed wet,
under tension to give a raised surface to the yarn.

Mixed constructions

All the elements that make up singles and plied yarn can be varied and mixed.
Fibres and colours, thickness and twist, singles can be different tensions

Using a spindle for extra thick yarn

and directions, they can change roles and have different additions. Bulky complicated yarns can be wound and plied by spindle if they are too bulky to pass through the eye of the flyer wheel.

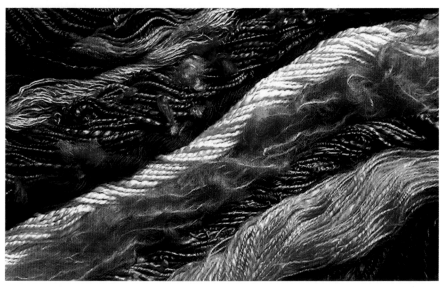

Yarns with mixed constructions. Hand spun by Prue Bramwell Davies, Nicky Hessenberg and Tricia Julian. Photograph: Ian Hessenberg

Mixed loop yarns. Photograph: Graham Murrell

Mixed bulky yarns loop wool scarf knitted by Diane Sullock

Yarn finishing

To set the twist permanently and to pre shrink yarns that might distort when woven into fabric, some yarns have to be put through a finishing process. Hand spinners often have to 'block' (damp and dry under tension) skeins of yarn to ensure an absolutely even balance that does not skew when used.

Scouring and setting

Woollen and worsted yarns are scoured with detergent. Lanolin is a by product of this process, which used to involve soaking in lime or dung. Worsted is heat set after weaving.

Singing or gassing

This process is traditionally the finishing for fine linens and some high lustre silks and sewing threads. Protruding anterior fibres are removed to produce a completely smooth high lustre yarn, usually after spinning and before plying.

Steaming

Steaming helps to set and bulk soft woollen yarns and ensures that subsequent dyeing is even.

Brushing and raising

Woollen and hair fibres can be brushed up in the yarn stage. The damp skeins are raised with soft wire brushes under tension. This procedure is used for knitted and woven fabrics that do not have the whole surface raised.

Waxing

Yarn can be passed through a waxing ring. This strengthens the yarn, and therefore causes fewer breakages in weaving and knitting. Synthetic fibres are passed through a waxing process to reduce pilling on the surface of the yarn.

Sizing

Sizing is usually starch based and makes cottons and linens more abrasion resistant.

Mercerising

Mercerising was invented in 1844. In it, cotton and linen yarn are stretched and treated with a solution of potassium hydroxide. This causes the fibres to swell and become smooth and round, becoming stronger, more lustrous and able to absorb dyes more easily.

Shrink proofing

Drying yarn under tension can give some resistance to shrinking. Wool can be drawn through a machine that removes the surface scales that interlock and cause shrinkage.

Testing

Commercial yarns are also put through a great many tests to check their performance under various sorts of stress and wear. Natural and manufactured yarns are tested for:

- and axial (longways and sideways) splitting
- levels of strain in bending and rotating
- recovery
- shrinkage
- static electricity and thermal conductivity
- flammability
- resistance to microbiological attack.

FIVE

Yarn in Fabric

How fabrics are affected by yarn design and structure

All textiles except felt are dependent on the yarn used to weave them for their character. The yarn creates the underlying construction that determines the behaviour and performance of the fabric. Fibre, twist, radius and elasticity as well as surface lustre, colour and spinning design make the yarn suitable for the particular sort of textile it is used to construct. Designers can use the yarns as a decorative feature of the fabric, or as the most suitable building units for woven designs, or to construct the best surface for a printed design. The yarn

Silk with Angora spots in knitted fabric. Photograph: Nigel Swift

Wensleydale gives a specific shape to the stitches in this scarf. Photograph: Graham Murrell. Scarf knitted by Mr M Designs

Fleece from the rare breed Wensleydale sheep at Stilereed farm

can be very visible, as in embroidery or tapestry, or can simply taken for granted and the whole textile seen as a woven or knitted structure.

Certain characteristics of yarn will follow a consistent pattern when used in any textile. In woven fabrics, loosely spun threads are more suitable for the weft, and threads with a higher twist per centimetre are suitable for the warp. Yarns spun from staple fibres will be more flexible and will drape better than yarns spun from filament fibres, which will give a stiffer fabric. Yarns used in knitted structures should be elastic and resilient for the fabric to keep its shape.

Combinations of fibres have been used as a design feature from ancient times. Remains of textiles show different fibres used in the warp and weft yarns, known as 'union' fabrics. The contrasting textures of fibre mixes can be used by the designer to create pattern effects. Wool/silk and linen/silk mixes, and increasingly combinations of viscose and manufactured fibres with unusual natural fibres such as soya and bamboo, make blends that combine the textures of both fibres.

The same woven or knitted pattern can be altered completely by changes in the twist of the yarn. The decorative quality of some woven designs, for example, damasks, depends entirely on changes in twist direction and the twist angle of the yarns used.

Contemporary textile artists have an enormous array of yarn designs and colours to choose from, with reflective and elastic qualities that have been developed from contemporary fibres and spinning techniques.

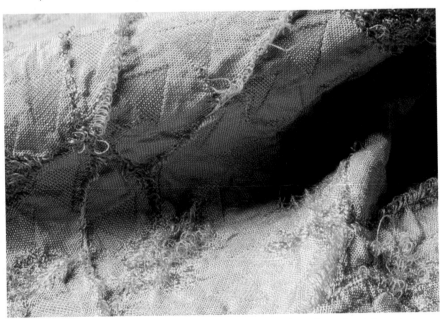

Weave designer Margo Selby uses contemporary yarns to give three dimensional drama to her fabrics. Photograph: Graham Murrelll

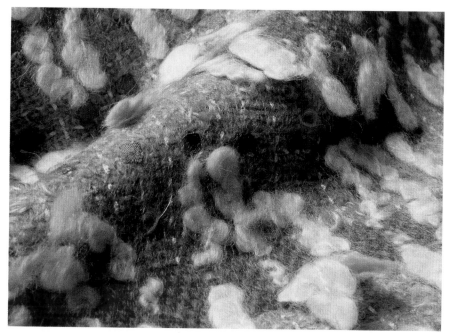

Eleanor Pritchard uses boucle and gimp yarns for contrasting textures.
Photograph: Graham Murrell

Yoko Hatakeyama, a design research fellow at the Ann Sutton Foundation uses different weights
of yarn in interchanging weaves. Photograph: Graham Murrell

Glossary of fabrics affected by yarn design and structure

The names of fabrics originally indicated the yarns and weaving patterns used to make them. Modern fabrics can sometimes imitate these traditional constructions, by copying the appearance and behaviour of the original designs, using machine made textures and manufactured fibres. Some of the more intricate fabrics are now only seen as costumes and textiles in museums.

Textiles in which the yarn plays an important role in the look and behaviour of the cloth

Barathea: Pre-war gentlemen's suit fabric. It has a minutely pebbled surface due to a mixed worsted and silk weft.

Bedford cord or *piqué*: Thick wadding threads make ridges in weft or warp, finer threads weave over them.

Brocade: Satin background using twisted silk thread and silk filaments in different colours for figuring and pattern.

Brocatelle: Brocaded pattern made by contrasting silk and linen threads.

Bombazine: Popular as a dress fabric in the 19th century. Traditionally woven from a tightly twisted silk warp and a worsted weft.

Calico: Medium thick unbleached cotton thread in the warp and weft, containing remnants of seed husk of cotton due to shortened fibre preparation.

Cambric: Smooth fabric now little used. Made from the longest line flax, tightly twisted into fine threads. Used in the 16th century for best quality collars and ruffs and later for corsets and handkerchiefs.

Chambray: Fabric with coloured threads in the warp and white in the weft.

Chenille: A yarn made from narrow woven strips with the weft cut to give a velvet surface. This gives the fabric its name.

Chiffon (silk): Made from reeled sheer continuous silk filament, woven in gum, with at least 30 ends per centimetre.

Chenille fabric. Photograph: Nigel Swift

Crepe: Pairs of alternate 'S'- and 'Z'-plied yarns in warp and weft. Contrasting torque of yarns means fabric is clinging and elastic.

Crepon: Crepe yarn in weft only, giving fluted vertical pleats.

Silk crepe showing pairs of alternate S and Z plied yarns. Photograph: Nigel Swift

Crepe de Chine: Filament silk warp and filament silk weft twisted into 4-ply crepe.

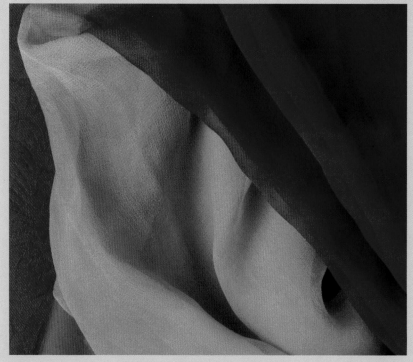

Crepe de Chine. Photograph: Nigel Swift

Moss crepe (silk): Woven from pairs of alternating 'S'- and 'Z'-ply crepe yarns in weft only to give a spongy handle and good draping quality.

Damask: Reversible figured fabric. The pattern is made by a mixture of satin and sateen weave in silk filament yarn to give lustrous surface. Some damasks use contrasting twist direction to accentuate pattern.

Denim: White weft yarn, coloured warp yarn with 2 up 1 down twill. Old fashioned cotton fabric with a corded effect from using two threads as one.

Dupion/Dupioni (silk): Made from irregular slubbed silk bassinette after the filament has been reeled off the cocoons.

Faille: Fabric with horizontal ribs because of thicker weft yarn. Ascending order of the prominence of weft ribs gives: taffeta, faille, grosgrain.

Fustian: Tightly spun worsted warp and cotton weft. Originally woven in the 17th century to imitate Indian palimpores, but British spinners could not spin cotton tightly enough for warp thread.

Gabardine: Worsted warp wool and cotton weft, the weft is impacted to give compact water-resistant fabric.

Georgette: Crepe twist yarns in plain weave. The yarns shrink in wet finishing, contracting the surface.

Hessian: Woven from thick jute yarn.

Jappe. Plain weave using continuous multiple filament silk threads.

Lawn: Woven from long staple fine spun cotton, plain weave.

Linsey-Wolsey: An archaic fabric with a linen warp and worsted weft.

Muslin: Plain weave in fine cotton 2 ply thread.

Organdy: Sized stiff cotton fabric used for curtains and dresses.

Organza: Sheer stiff cloth woven from filament silk tightly twisted and wrapped 2 ply with I ply retarded. Woven in gum.

Mousseline: Plain weave in fine silk filament. Weft in 2 ply S and Z mixed to give slightly textured surface.

Pongee or shantung: Woven irregular wild or Tussah silk.

Poplin: Originally silk warp and worsted weft with more warp ends than weft picks. Later 2 fold cotton replaced worsted.

Ratine: Loosely woven cloth using textured yarns.

Repp: Weft rib cloth. Thick low twist yarns packed into weft.

Satin: Woven as continuously broken twill with floats creating a reflective surface. Can be warp- or weft-faced, and the back of either type will have shorter floats and therefore be less reflective. Duchess satin has more ends and picks and is woven from finer flatted silk filament.

Silk, sateen and duchess satin. Photograph: Nigel Swift

Seersucker: Alternate thick and thin ends and picks of cotton yarn. Thin threads are high twist so that fabric cockles when taken off the loom and wetted to shrink high twist threads further. Now can be woven using unset nylon which shrinks when finished in hot water and can be set again.

Serge: 2 up 2 down twill weave worsted. Cheaper serge can be worsted warp and linen weft but this develops a shine.

Spandex (Lycra): This is made from yarn in which the polymer chains have exceptionally stretchy areas and stable areas that hold the chain together.

Taffeta: A plain woven fabric with a higher twist in the weft, forming ribs.

Tweeds: Heavy weight fabric with colour effects from mixed woollen yarns. Bannockburn tweed is woven from Cheviot wool (2 ply). Donegal tweed has brightly coloured spots incorporated in weft yarn. Harris tweed is Cheviot wool yarn with kemp hairs (white) left in.

Velvet: Extra weft or warp of high density low twist silk to form pile. Background warp and weft, high twist silk.

SIX

Contemporary Yarns

The new generation of yarns shown at international design fairs across the world seem to be the product of a second industrial revolution in textile production. Huge global companies are constantly developing new uses for textiles and new fashions in wearing them, and the yarn production industry employs advanced technical expertise and sophisticated processing techniques to make innovative yarns that are the basic units of these 21st century textiles. Yarn production is no longer based on traditional manufacturing, but is a global industry driven by scientific innovation from institutions such as the University of Manchester Institute of Science and Technology.

The most interesting and specialised yarns are shown at Expofil in Paris where manufacturers collections are shown twice a year, in February and September. Fancy and classic yarns, mostly for the fashion knitwear market are shown in two collections in Florence, Italy at Pitti Immagine Filati.

Expofil Paris

Fibre producers and spinners are all classified as 'yarn makers' and they employ advanced technology to produce and spin innovative fibres from some surprising sources.

New fibres

Some of the most interesting new fibres are the adaptable micro- or nanofibres. Cut longitudinally into hair-like mono-filaments from an acrylic thread, they can be coated with ceramic composites for sportswear, or coated with heat insulating silk-like material. Electrostatic fibre-producing machines have now been developed that can produce filaments of extremely small diameter, but great strength, for example Kevlar which is soft and stable and can be as fine as 0.9 denier.

Aquacel™ or aquator is derived from seaweed; the cotton-like fibres can absorb 50 times their weight in water and draw moisture from the skin, so are of use for sportswear. New nylon threads from Du Pont™ have a high melting point and so are fire resistant. They can be thermo-fixed, so nylon tights can hold their shape. Elastane™/spandex (trade name Lycra™) is now an established fibre that can increase stretch and textile recovery. It can be blended with natural fibres such as cotton and wool or with acrylic fibres.

The Italian company Bugetti Filati, have developed two qualities of stretch steel that is suitable for clothing. Although the plastic from mineral water bottles has been used for carpet yarn for some years, it has now been developed into a yarn fine enough to be blended with Angora rabbit hair for gloves and scarves, made by Rhoyd Eco.

Lenpur™ or 'vegetable cashmere' is a soft delicate fibre made of cellulose from the sustainable white pine. The recent move to produce yarns from more

Bamboo, nettle and hemp fibre. Photograph: Graham Murrell

Innovative yarns mixing natural fibres.
Filpucci of Italy

sustainable and ecologically con-scious sources has seen a develop-ment in some vegetable mixes of bamboo, soya, nettle and paper.

The Turkish company Sanko is now an established producer of organic cotton fibre, and Mario Boscelli has produced a range of silk and soya yarns. Frisotine™ is a mix of viscose and nettle made by Saint Lievin. The same movement has seen a significant rise in the producers of organic cotton and wool, and the fashion for a more natural and rustic effect has seen the develop-ment and blending of mixes of natural fibres, for example, shrink-resistant wool, rot-resistant cotton, and blends of wool and bamboo, or cashmere and nylon.

Filatura di Grignasco has produced a blend of Tactel™ and Merino which is soft and shiny enough for knitwear.

Some of the innovations are for medical or industrial use. Sicofil's fluorescent yarn recharges in light and several antibacterial and antiseptic fibres have been adapted from vegetable mixes for use in hospitals, both as clothing and for use in medical procedures. Biokryl™, Courtauld's antibacterial acrylic fibre, which can slow-release chemicals, also has a clothing application for sports socks.

Contemporary yarn spinning

Contemporary yarns can be produced in great quantities, at great speed and in innumerable designs by two methods: machine spun yarns and extruded filament yarns.

Machine spun yarns

Innovative contemporary yarns can be machine spun from natural fibres or mixtures of natural and manufactured fibres.

Condenser spinning and draw spinning

These can produce slubbed yarn by incorporating nepps or slubs of uncarded fibres into the pre-spinning drafting process. Intermittent acceleration of the machine forms slubs in the yarn made from a mixture of long and short fibres.

Fancy ring spinning

Special effect yarns can be spun by a specialised ring spinning machine, in one operation. A thick slubbed yarn can be fed into the ply of two supporting yarns at a faster rate, lowered in between the plies in an S shape to form a

Modern machinery for winding fancy yarns. Filpucci of Italy

S shaped gimp. Rowan yarns cotton braid. Mix of cotton 68 per cent viscose 22 per cent and linen, soft knitting yarn

gimp, or fed in extra quickly to form loops. Machines can twist several threads together to form ruffles and snarls, mixing in thicker and thinner threads.

Extruded filament yarns

A great variety of contemporary extruded filament yarns are now available to the textile industry. Although expensive to set up, the machinery can produce large quantities of specialised yarns at great speed.

Jet interlacing

Invented in the 1980s by Du Pont, jet interlacing means that the filaments are passed through an air jet and simultaneously twisted and interlaced or entangled.

Crimped viscose

To produce crimped viscose, during the extrusion process, a thicker skin is formed down one side of the filament, this pulls the thread into crimps, giving it a softer, warmer handle than viscose.

Air bulked yarns

Bi-component yarns, made of two fibres which are blended or processed together, are put through a heat process to contract one fibre and bulk up the yarn. This process is usually used to treat acrylic fibres.

Rowan yarns blended with viscose to give the glamorous effect of shimmering metal thread

Aerated yarns

The manufactured filament can be aerated with bubbles to create textured knitted fabrics.

De-knitted yarn

Filament polyester and nylon can be extruded, knitted, heat-set and then unravelled to make a kinked, textured yarn.

Multiple filament yarn

The spinnarets have many holes so can extrude multiple filaments which can be cut up and spun together to create multiple filament yarn.

Electrostatic spinning

This is the latest method of filament spinning, developed in the USA. The filament is extruded by a jet from a nozzle, and stopped at the point it becomes unstable. The flow rate, fluid density and the collector distance dictate the radius of the fibre, which can be so small it has to be measured by an electron microscope.

The new generation of yarns

This new generation of machine made yarn is constantly innovative, allowing contemporary textiles to break the barriers of what is possible in terms of elasticity, heat resistance and machine washability.

Organic cotton from Peru. Image from
Handknitting.com

Hemp yarn for knitting. Image from
Handknitting.com

Eyelash yarn

Yarn stocked by Handknitting.com, run from Prince Edward County in Canada, includes eco-organic cotton from Peru and also hemp yarn, which could possibly replace some synthetics, as it yields more fibre than any other plant, and does not exhaust the soil in which it grows.

Fashion yarns include 'eyelash yarn' and other fantasy effect yarns for contemporary hand knitting.

The Nuno Corporation of Japan have designed and made some of the most innovative yarns and textiles of the past two decades. Some of their most interesting developments have been threads with incompatible shrink ratios. These are intertwined and put into a hot dryer to be made into the sculptural

Nuno Corporation Japan, Banana fibre

Metallic film bonded onto yarn filaments then melted. Nuno Corporation, Japan

Conductive yarns are used to enable panic alarms and talking modules to be woven on to the electronically active fabrics. Photo from Intelligent Textiles Department of Design, Brunel University

textiles popular with Japanese fashion designers. Other innovations include Okinawa banana fibre-coated cottons, and a range of synthetics finished in different ways to give dramatic effects, for example, metallic films are bonded to the filaments which are then melted to create a transparent filigree.

Bridging the fashion and technical markets 'Intelligent Textiles' of Brunel University produce conductive yarns which are woven into sensory textiles for medical applications. These sensory 'smart' fabrics may be developed for the domestic, fashion and sports markets as the yarns of the future.

Bibliography

Aspin, Chris, *The Woollen Industry*, Shire, 2000
Baines, Patricia, *Flax and Linen*, Shire, 1985
Baines, Patricia, *Spinning Wheels*, Batsford, 1977
Bush, Sarah, *The Silk Industry*, Shire, 1987
Chadwick, Eileen, *The Craft of Handspinning*, Batsford, 1980
Corbman, Bernard, *Textiles: Fiber to Fabric*, McGraw-Hill, 1983
Emery, I., *The Primary Structures of Fabrics*, Thames and Hudson, 1994
Giejer, Agnes, *A History of Textile Art*, WS Maney & Son
Hulton, Mary, *True as Coventry Blue*, Coventry Branch of the Historical Association, 1995
Kroll, Carol, *The Whole Craft of Spinning*, Dover Publications, 1981
Leadbeater, Eliza, *Spinning and Spinning Wheels*, Shire, 1979
Lewis, June, *From Fleece to Fabric*, Robert Hale, 1983
Maik, Jerzy, *The Textiles of Pomerania in the Roman Period and in The Middle Ages*, Wydwnictwo Polskiej Akademii Nauk, 1988
Miller, E., *Textiles; Properties and Behaviour in Clothing Use*, Batsford, 1988
Ponting, K.G., *Drawings of Textile Machines - Leonardo Da Vinci*, Humanities Press, 1979
Ponting, K.G. and Jenkins, D., *The British Wool Textile Industry 1770-1914*, Scolar Press, 1987
Rose, Mary, *The Lancashire Cotton Industry*, Lancashire County Books, 1996
Ross, Mabel, *The Essentials of Yarn Design*, Mabel Ross, 1983
Ross, Mabel, *Handspinners Workbook - Fancy Yarns*, Mabel Ross, 1989
Scott, Philippa, *The Book of Silk*, Thames and Hudson, 1993
Swanson, Heather, *Medieval Artisans*, Blackwell, 1989
Taylor, M.I., *Technology of Textile Properties*, Forbes Publications, 1990
Varney, Diane, *Spinning Designer Yarns*, Interweave Press, 1987
Walton-Rogers, P., Bender-Jorgensen, I. and Rast-Eicher, A., *The Roman Textile Industry and its Influence*, Oxbow Books, 2001
Wild, John Peter, *Textiles in Archeology*, Shire, 1988

Suppliers and Places of Interest

UK Suppliers

Blueberry Angoras
Ffynnon Watty
Moylegrove
Pembrokeshire SA43 3BU
Tel: 01239 881 668
Fibres, fleece and dyes.
Weaving/spinning/dyeing equipment

Devon Goat Co.
Westcott Farm, Oakford, Tiverton
Devon EX16 9EZ
Tel: 01398 351 173
www.devongoat.co.uk

Fibres and Fleece, Fibrecrafts
Old Portsmouth Road
Peasmarsh, Guildford, Surrey GU3 1LZ
Tel: 01483 565 800
sales@fibrecrafts.com
www.fibrecrafts.com
Fibres, fleece and dyes.
Weaving/spinning/dyeing equipment

Hilltop Spinning and Weaving Centre
Windmill Cross, Canterbury Road
Lyminge, Folkestone, Kent CT18 8HD
Tel: 01303 862 617
www.handspin.co.uk
Fibres, fleece and dyes.
Weaving/spinning/dyeing equipment.
Books, craft kits, videos

Jameson and Smith
Shetland Wool Brokers (mail order)
90 North Road
Lerwick
Shetland Islands UK
www.shetland-wool-
brokers.zetnet.co.uk/

Rosemary Kitchingman/Raw Fibres
The Old Signal Box, Station Workshop
Robin Hood's Bay
N.Yorks YO22 4RA
Tel: 01947 880 632
Wool, silk, mohair, alpaca, cotton, flax,
soya, bamboo fibres, Shetland fleece.

Meon Valley Alpacas
Grooms Cottage
Midlington Hill, Droxford
Hants SO32 3PY
Tel: 01489 878 663
www.mvalpacas.co.uk
Fine alpaca fleeces, yarn, all colours.

Michael Williams
74 Norfolk Road, Sheffield S2 2SY
Tel: 01142 721 039
Handcrafted spinning and weaving
equipment

Moondance Wools
Springhill Farm, Coldingham Moor
Berwickshire, Scotland TD14 5TX
Tel: 018907 71541
www.moondancewools.com
Tuition, fleece, fibres and yarn

P&M Woolcraft
Pindon End, Hanslope
Milton Keynes MK19 7HN
Tel: 01908 510 277
www.pmwoolcraft.co.uk
Ashford, Lendrum, Louet wheels and
equipment

Sue Hiley Harris
90 The Struet, Brecon, Powys LD3 7LS
Tel: 01874 610 892
www.suehileyharris.co.uk
Silk fibres and yarn

Teal & Co
Woodcross Farm, Foxholes Lane
Creech St Michael, Taunton TA3 5EF
Tel: 01823 442 880
Hand woolcombs, tuition and books

Wingham Woolwork
70 Main Street, Wentworth
Rotherham, S.Yorkshire S62 7TN
Tel: 01226 742 926
Fibres and spinning equipment

UK Places of Interest

Recommended by Association of
Guilds of Weavers, Spinners and
Dyers (www.wsd.org.uk)

Bankfield Museum and Art Gallery
Boothtown Road
Halifax, West Yorks HX3 6HG
Tel: 01422 354 823

Costumes and Textiles
Coldharbour Mill
Cullompton
Devon EX15 3EE. Tel: 01884 840 858
www.coldarbourmill.org.uk
Demonstrations: carding, spinning,
weaving

Hatfield House
Hertfordshire AL9 5NQ
Tel: 01707 287010
www.hatfield-house.co.uk
Tapestries

Knole House
Sevenoaks, Kent TN15 ORP.
Tel: 01732 462 199
Flemish tapestries, C17 furnishing fabrics

Laxey Woollen Mills, Glen Road
Isle of Man, Laxey
Tel: 01624 861 395
Weaving looms operating

Lavenham Guild Hall
Market Place, Lavenham
Sudbury CO10 9QZ
Tel: 01787 247 646
lavenhamguildhall@nationaltrust.org.uk
Medieval wool trade hall

Museum of Costume
Bennett Street
Bath BA1 2QH
Tel: 01225 477 173
costume-bookings@bathnew.gov.uk
www.museumofcostume.co.uk

Pitt Rivers Museum
South Parks Road, Oxford
Tel: 01865 270 927
www.prm.ox.ac.uk

Snowshill Manor
Gloucestershire WR12 7JU
Tel: 01386 842 814
snowshillmanor@nationaltrust.org.uk
Collection of spinning wheels

US Suppliers

Apple Hollow Fiber Arts
732 Jefferson Street
Sturgeon Bay, Wisconsin WI 54235
Tel: 888 324 8302 or 920 746 7815
www.applehollow.com
Fibres and equipment

Artfibers
124 Sutter St
San Francisco, CA 94104
Tel: 888 326 1112 (Toll free)
www.yarndesign.com
Yarn fiber mill and knitting lounge

Craft Yarn Council of America
PO Box 9, Gastonia NC 28053
Tel: 704 824 7838
www.craftyarncouncil.com

Fiber Crafts
38 Center St, Clinton
New Jersey 08809
Spinning supplies and books

Foxglove Fiberarts Supply
8040 NE Day Road West, Suite 4F
Bainbridge Island, WA 98110
Tel: 206 780 2747
www.foxglovefiber.com
sales@foxglovefiber.com
Ashford spinning wheels, looms.

Weaving Workshop
920 East Johnson
Madison WI 53703
Tel: 608 255 1066
http://new.enterit.com/Weaving1066/

The Woolery
PO Box 468, Murfreesboro,
NC 27855
Tel: 800 441 9665
www.woolery.com
Equipment and supplies for
spinning and fibre preparation

US Places of Interest

American Textile History Museum
491 Dutton Street
Lowell, MA 01854-4221
Tel: 978 441 0400
www.athm.org

Black Sheep
Handspinner's Guild
Varna Community Center
Route 366, Ithaca, NY
www.lightlink.com/devine/BlackSheep

Handweaver's Guild of America (HGA)
1255 Buford Highway, Suite 211
Suwanee, Georgia 30024
Tel: 678 730 0010
hga@weavespindye.org
www.weavespindye.org

The Hudson River Museum
511 Warburton Avenue
Yonkers, NY 10701
Tel: 914 963 4550
www.hrm.org
Textile exhibits

Navajo-Churro Sheep Association
www.navajo-churrosheep.com/

Spin-Off Magazine
Interweave Press
PO Box 469115, Escondido
CA 92046-9115
Tel: 800 767 9638
SpinOff@interweave.com
www.interweave.com/spin

The Windham Textile and History
Museum (Mill Museum)
157 Union St and Main St
Willimantic, CT 06226
Tel: 860-456-2178
www.millmuseum.org
Heritage of American textile industry

Index